Fertilization

CW00675540

The Beginning of Life

Fertilization

The Beginning of Life

Brian Dale
Centre for Assisted Fertilization, Naples

CAMBRIDGE
UNIVERSITY PRESS

University Printing House, Cambridge CB2 8BS, United Kingdom

One Liberty Plaza, 20th Floor, New York, NY 10006, USA

477 Williamstown Road, Port Melbourne, VIC 3207, Australia

314–321, 3rd Floor, Plot 3, Splendor Forum, Jasola District Centre, New Delhi – 110025, India

79 Anson Road, #06–04/06, Singapore 079906

Cambridge University Press is part of the University of Cambridge.

It furthers the University's mission by disseminating knowledge in the pursuit of education, learning, and research at the highest international levels of excellence.

www.cambridge.org
Information on this title: www.cambridge.org/9781316607893
DOI: 10.1017/9781316650318

© Brian Dale 2018

This publication is in copyright. Subject to statutory exception and to the provisions of relevant collective licensing agreements, no reproduction of any part may take place without the written permission of Cambridge University Press.

First published 2018

Printed in the United Kingdom by TJ International Ltd. Padstow Cornwall

A catalogue record for this publication is available from the British Library.

Library of Congress Cataloging-in-Publication Data
Names: Dale, Brian, author.
Title: Fertilization : the beginning of life / Brian Dale.
Description: Cambridge, United Kingdom ; New York, NY : Cambridge University Press, 2018. | Includes bibliographical references.
Identifiers: LCCN 2018010116 | ISBN 9781316607893 (paperback)
Subjects: LCSH: Fertilization (Biology) | Germ cells. | Gametogenesis. | MESH: Fertilization – physiology | Germ Cells – physiology | Gametogenesis – physiology | Embryonic Development
Classification: LCC QH485 .D35 2018 | NLM QH 485 | DDC 571.8/64–dc23
LC record available at https://lccn.loc.gov/2018010116

ISBN 978-1-316-60789-3 Paperback

Cambridge University Press has no responsibility for the persistence or accuracy of URLs for external or third-party internet websites referred to in this publication and does not guarantee that any content on such websites is, or will remain, accurate or appropriate.

Every effort has been made in preparing this book to provide accurate and up-to-date information that is in accord with accepted standards and practice at the time of publication. Although case histories are drawn from actual cases, every effort has been made to disguise the identities of the individuals involved. Nevertheless, the authors, editors, and publishers can make no warranties that the information contained herein is totally free from error, not least because clinical standards are constantly changing through research and regulation. The authors, editors, and publishers therefore disclaim all liability for direct or consequential damages resulting from the use of material contained in this book. Readers are strongly advised to pay careful attention to information provided by the manufacturer of any drugs or equipment that they plan to use.

This book is dedicated to Loredana, Daniela, Peter, Roberta and Rebecca.

Contents

Preface

Despite its importance in human-assisted reproduction, agriculture and fisheries, not to mention our very existence, the subject of fertilization receives little attention in textbooks of reproductive or developmental biology. The aim of this book is to introduce the reader to this fascinating process where two highly specialized cells interact to form a new life. Using examples from the echinoderms, ascidians, amphibians, fish, mammals and other phyla, I will try to show that, despite the variability in form of metazoan gametes, the mechanism of fertilization is highly conserved throughout the animal kingdom. Since there are over 3,000 specialized papers published annually on this subject, I can only outline the basic principles involved in fertilization and invite readers who require more details to refer to more specialized texts. Fertilization is about the transformation of a quiescent oocyte, which is primarily concerned with attracting its partner gamete, into a dynamic zygote that undergoes a cascade of predetermined activation events to set the scene for the early embryo. Many events are about reorganizing components laid down in the oocyte during oogenesis; therefore, I have summarized the principles of producing gametes in Chapter 2. In some animals this maternal information is polarized in the oocyte, and partitioning at cleavage gives rise to different cell lineages in the early embryo, whereas in others, early blastomeres remain totipotent. These differences are covered in Chapter 7. The main theme of this book is to decipher the spatial and temporal complexity of fertilization under natural conditions and, in particular, to show how each gamete induces successive physiological changes in its partner that are essential for the formation of the zygote. Fertilization studies are fraught with contrasting ideas, often arising from artefacts induced by in vitro studies, where techniques to prepare gametes have distorted the results of experiments. For example, techniques that bypass essential stages of the fertilization process, such as removing extracellular coats or avoiding gamete fusion by micro-injecting sperm into oocytes, not only change the physiology of the cells but also give the impression that these structures or processes are not required for fertilization. Chapter 6 (modified from *The Encyclopedia of Reproduction*, second edition, Elsevier) looks at the dynamics of sperm–oocyte interaction, taking care to reflect the situation under natural conditions and avoiding misleading interpretations from laboratory experiments. To improve the flow of the book, basic biological processes common to all cells, such as cell division, electrical properties and metabolism are treated in a separate chapter at the end of the book.

I have had the honour to work with some outstanding scientists – Alberto Monroy, Giuseppina Ortolani, Berndt Hagstrom, Louis DeFelice, Yves Menezo and Jacques Cohen, to mention a few – and the privilege to work in a period of unparalleled enthusiasm, where scientists such as Alberto Monroy, Daniel Mazia, Jean Brachet, Eric Davidson, Mike Bedford, Robert Edwards and Ryuzo Yanagimachi set the pace in studies of fertilization. I wish to thank the Stazione Zoologica in Naples, the first and most unique marine biology laboratory in the world, for allowing me the opportunity to work there from 1976 until 1999 on a variety of invertebrate and vertebrate species, and the late Alberto Monroy for introducing me to this fascinating subject.

Chapter 1

Introduction

The history of fertilization is as fascinating as the subject itself, stretching from Greek philosophers through medieval times to the present day, raising many passionate controversies between scientists and philosophers that are often tainted with religious beliefs. Advances were due in part to the invention of scientific instruments, such as the light microscope of Leeuwenhoek, but the main thrust has been the intuitive curiosity of some outstanding scientists. Although diagnostic tools are different today, the gametes are not. We can rest assured that, when observing gametes, what passes through our imagination captured that of our predecessors and to them we should always give due credit.

Hippocrates (460–370 BC) argued that both male and female 'semen' existed and that these mixed in the uterus to form an embryo. Aristotle (384–322 BC) considered animals to be divided into two groups: the bloodless kind, such as insects which generated spontaneously, and all the others that had to mate in order to reproduce. Aristotle favoured a male-centred view where, although the female provided the matter through her menstrual blood, the male semen gave form to the matter. The ideas of Aristotle and Hippocrates dominated thought in the Western world for over 1,500 years until the English physician William Harvey (1578–1657) published his landmark book *De Generatione Animalium* in 1651. The frontispiece of the book depicts Zeus holding two halves of an egg inscribed with the words *ex ovo omnia,* with plants, insects, fish, amphibians, reptiles, birds and mammals emerging from the shell. The concept 'Everything comes from the egg' gained ground over the next 25 years and was supported by the observations of Francesco Redi in 1668 and J. Swammerdam in 1669, both of whom worked on insects. Reinier de Graaf in 1672, provided a detailed account of the human female reproductive tract and, from studies mainly on the rabbit, suggested that ovarian follicles were in fact eggs that were found in the fallopian tubes after copulation. It was not until the early nineteenth century (1827) that the Estonian Karl Ernst von Baer actually observed the mammalian oocyte under the microscope and illustrated the oocyte lying in the Graafian follicle of the ovary of a sow.

In 1677, a Dutch draper, Antonie van Leeuwenhoek, who had begun making simple single-lens microscopes, observed tiny animalcules in his own semen that later were given the name 'spermatozoa' (which translates as 'semen animals'). Leeuwenhoek did not immediately grasp the importance of the discovery of spermatozoa and thought they were another example of animalcules which were found in other biological material, including pus cells. He did however, in 1699, fight the notion that the spermatozoon contained a preformed human, the homunculus, and concluded that there were two sorts of animalcules, one female and one male.

Lazzaro Spallanzani, an Italian priest, published his 1785 classical work *Expériences pour server a l'histoire de la génération des Animaux et des Plantes* in Geneva and was the first person to successfully carry out artificial insemination. He also investigated the effect of temperature and certain chemicals on the fertilizing power of spermatozoa from amphibians. For example, he showed that toad sperm lost its fecundity after six hours at 70°F, but remained fertile for up to 25 hours if it was kept in an icebox at 40°F (the forerunner of today's cryobiology). As a consequence of this latter work, he tried to trigger development with a variety of chemical agents; in fact he introduced the first experiments on artificial parthenogenesis. Although he is often quoted as the scientist who promoted the idea of the activating capacity of spermatozoa, he actually held the opposing idea. In experiments, he showed that if seminal fluid was filtered through filter paper, the filtrate had no fertilizing power, whereas the residue would fertilize. He concluded wrongly that filtration removes the fertilizing power of seminal fluid and sustained that the fertilizing power must remain on the filter paper. Spallanzani missed the fundamental conclusion that the spermatozoa themselves were the fertilizing agents.

In the first half of the nineteenth century, two main fertilization theories circulated. In 1824, Prevost and Dumas proposed that the spermatozoon actually penetrated into the egg, while Bischoff in 1841 supported the idea that the spermatozoon acted by contact only. George Newport in 1853 showed in amphibians that it was the spermatozoon *not* the seminal fluid that 'was the sole agent of impregnation' (p. 231). After long years of study in which he was adamant that the spermatozoa penetrated the layers of the oocyte, he finally detected spermatozoa in the oocyte's 'yelk'; the oocyte cytoplasm. However, Newport did not directly observe penetration of the spermatozoon and believed that many spermatozoa were required for fertilization, stating 'Fecundation is not the result of a single isolated spermatozoon' and 'a plurality of spermatozoa is necessary for the full impregnation of the egg and the production of a robust and healthy embryo' (1853, p. 245). Newport's work was fundamental in showing that gamete function was dependent on time after ovulation or ejaculation and was also temperature dependent. He also describes the first activation event in amphibian oocytes – that is, the formation of a space or 'chamber' between the vitelline envelope and 'yelk' within 90 minutes of activation and showed that this was in fact located at the animal pole where sperm penetration was preferential. Around the same period, others also maintained they had observed penetration of the ovum by the spermatozoon, such as Barry in 1840, Meissner in 1855 and Keber in 1854, who placed special emphasis on the micropyle seen in some animals as an adaptation for the entrance of a spermatozoon.

The first direct evidence for sperm penetration was not made until 1879 by the Swiss zoologist Hermann Fol using the starfish *Asterias*. Fol observed a thin filament extending from the spermatozoon through the jelly layer of the oocyte to the oocyte's surface. Although Fol dismissed the idea that the filament arose from the spermatozoon itself, he alluded to the fact that the filament pulled the spermatozoon to the oocyte's surface. He also observed that, as the spermatozoon moved through the jelly, a protrusion from the oocyte's surface, the fertilization cone, appeared to rise and meet the oocyte. Oskar Hertwig in 1875, taking advantage of the remarkable clarity of sea urchin oocytes, described one of the fundamental phenomena of fertilization, the sperm nucleus and its aster with the approach of the sperm nucleus to the female nucleus and their apparent fusion. In 1883, Van Beneden, in his classical paper on the parasitic nematode worm *Ascaris*, showed that the pronuclei do not unite but are included in a single amphiaster and that each pronucleus produces two chromosomes. He thus demonstrated for the first time that there are equal numbers of male and female elements in the nuclei of the early embryo.

Theodor Boveri in 1887–1888, again using *Ascaris*, stated 'the egg is devoid of the organ of cell division, the centrosome; capacity for division, hence the initiation of the developmental processes, is restored through the introduction of a centrosome into the egg by the spermatozoon' (see Lillie 1916, p. 48). The paternal control of cell division was thus introduced. In the Stazione Zoologica in Naples in 1888, Boveri, now using the sea urchin, not only promoted his theory on the role of the centrosome in fertilization and early development, but he also discovered the jelly canal that marks the animal pole and showed that 'normal development is dependent on the normal combination of chromosomes and this can only mean that the individual chromosomes must possess different qualities' (see Lillie 1916, p. 48). Later in 1901, Boveri observed a ring of pigment in the oocyte of the sea urchin *Paracentrotus lividus* and related the polarity of the larva and cleavage to this equatorial ring. From this he recognized the importance of the vegetal cytoplasm and the micromeres 'that the area nearest the vegetal pole possesses the greatest potential to bring development to the pluteus stage' (see Ernst 1997, p. 253). These observations were clearly the precursor to the concept of the organizing centre forwarded by his future student Hans Spemann in 1924.

Thus by the end of the nineteenth century the morphological analysis of fertilization was fairly complete. Shortly afterwards scientists attempted to imitate the action of the spermatozoon by chemical and physical agencies. The scientists of the day coined the term 'irritable protoplasm' to describe the ease with which the oocyte surface could be altered. Embryonic surface waves, although previously noted by Fol in 1887, were first described by E. Conklin in 1905 in the ascidian *Cynthia partita*. These, in fact, were the mechanical manifestations of what we now know as the calcium waves that are generated in all oocytes at fertilization. In 1919, Ernest Everett Just, the first black American scientist, showed that 'before the actual elevation of the fertilization membrane, some cortical change beginning at the point of sperm entry sweeps over the egg, immunizing it to other sperm'. In 1939, in his landmark book, Just suggests this change may be attributable to nerve conduction, 'because among animal cells it is the most highly excitable and the most rapidly conducting' (p. 114).

Frank Lillie in 1916 introduced the quantitative aspect of fertilization, noting that the reaction may exhibit varying degrees of incompleteness. Lillie also states a fundamental rule in fertilization, that is, the spermatozoon will not fertilize until it is fully differentiated. Jacques Loeb, at the beginning of the twentieth century, showed that ion concentration and type were important for fertilization, starting the trend of chemical embryology, and in 1913 he successfully activated sea urchin eggs with butyric acid, resulting in normal cleavage and complete parthenogenetic development.

The undisputed innovator of experimental embryology, Sven Horstadius, actively published in the field for over 50 years from the 1920s to the 1970s. He created the first fate map of early sea urchin development and is known for his blastomere isolation and transplantation experiments. He showed that the entire embryo did not form without cells from the vegetal region. Advances in the detection of nucleic acids led to Jean Brachet showing in 1933 that sea urchin eggs must contain both DNA and RNA, and he came to the conclusion that nucleic acids must take part in the synthesis of proteins. Brachet sustained that the sea urchin oocyte was as an ideal organism for the study of this new area of molecular biology, well before the discovery of the structure of DNA by Watson and Crick in 1953. Finally, with the advances in electron microscopy in the 1950s and 1960s, A. Colwin and L. Colwin from the United States and J. Dan from Japan painstakingly described the various stages of the acrosome reaction in many invertebrates from starfish to polychaetes, pinpointing this reaction as the key to successful sperm–oocyte interaction.

Medical and veterinary interests promoted research in mammalian fertilization in the early twentieth century when the Russian School of Ivanov developed artificial vaginas and insemination techniques to be used in horses, cattle and sheep (surprisingly over a hundred years after the first application of the technique by Spallanzani). Since then, the use of artificial insemination techniques has progressed rapidly until the present day. Now, techniques of in vitro fertilization (IVF), embryo culture, cryopreservation techniques and genetic assessment of gametes and embryos are widely applied throughout the world. A major leap forward in this technology was made in the late 1950s, when the team led by Chris Polge in Cambridge, England, developed techniques to freeze and store animal spermatozoa (some 200 years after the discovery of Spallanzani). This same period of time also saw the development of methods to isolate and manipulate the female gamete. In vitro maturation of mammalian oocytes was first reported by Pincus in 1935, when it was observed that the primary oocyte of the rabbit resumed meiosis spontaneously when liberated from its follicle and placed in a suitable culture medium. It was not, however, until 1968 that Joe Sreenan in Ireland observed in vitro nuclear maturation in bovine oocytes recovered from slaughterhouse cattle. The present day IVF technology derives from the birth in 1978 of Louise Brown in the UK from human embryos produced in vitro (Steptoe and Edwards 1978). While, assisted reproductive technologies (ART) have been developed primarily to alleviate sterility, the possibility of having human gametes in vitro has led to a surge in pure research on human gametes and fertilization.

Chapter 2

Producing Gametes

Although much of our information on fertilization has come from studies on the invertebrates, which account for 95 per cent of animal species, vertebrate models have been instrumental in our knowledge of gametogenesis. The first phase in the sexual reproduction of animals is gametogenesis, a process of transformation whereby certain cells become the highly specialized sex cells, called oogenesis in the female and spermatogenesis in the male. In both sexes, gamete production may be divided into three phases: mitotic proliferation, meiotic division and differentiation. First, the primary gametes need to multiply in number, and then the chromosome number needs to be halved. Finally cytoplasmic modifications, involving massive growth in the oocyte and reduction in volume in the spermatozoon, prepare the gametes for their imminent interaction. Gametes are produced in great quantities in most animals. For example, a girl at puberty may have 300,000 immature oocytes which may remain in a dormant state in her ovaries for up to 50 years, while the human testis produces up to 500 spermatozoa per gram of testis per second throughout most of a male's adult life. In the male, each primary spermatocyte divides meiotically to produce four spermatids, each destined to become a functional spermatozoon: in the female, of the four cells produced from each primary oocyte, only one develops into a viable oocyte (Figure 2.1). An unequal distribution of cytoplasm at division results in the production of three small cells, the polar bodies, which eventually degenerate. A further distinction is that the spermatozoon acquires the ability to fertilize after meiosis is complete, while the oocyte interacts with the spermatozoon before the completion of meiosis. Exceptions are the sea urchins and some coelenterates where the oocytes have completed meiosis before fertilization. The process, called cytoplasmic maturation by Delage in 1901, whereby the oocyte attains the ability to interact with the spermatozoon is independent of the nuclear maturity. In oocytes that are normally fertilized before the completion of meiosis, the male nucleus remains quiescent in the cytoplasm until meiosis is complete.

General Considerations of Gametogenesis

Oocyte Growth

The growth of oocytes usually takes place over a relatively long period of time, and the increase in size is dramatic. The frog oocyte is an extreme example of this. The young oocyte, with a diameter of less than 50 μm, grows over a period of three years to reach a final diameter of 1500 μm – this represents an increase in size by a factor of over 20,000. Mammalian oocytes are much smaller, with a time scale of weeks rather than years in their growth period. However, the increase in size is also considerable. For example, the mouse oocyte grows from some 20 μm to a final diameter of 70 μm, an increase in volume by a factor of 40-fold. All oocytes are large, certainly larger than the average somatic cell, which is usually about 10 μm in diameter. The size of the full-grown oocyte depends principally on the amount of stored foodstuffs in the cytoplasm, which probably depends also on the mode of development of the fertilized embryo: for example, the frog oocyte must support development to the tadpole, a large structure capable of swimming (four to five days post fertilization), whereas the mammalian oocyte must support development to form a much smaller structure, the blastocyst which has implantation as its unique goal (six to seven days post fertilization). The nucleus of oocytes also enlarges and is called the germinal vesicle. Yolk is the major food-storage product, although large quantities of lipid and glycogen granules are also found in some oocytes. The chemical composition of yolk varies from species to species according to its

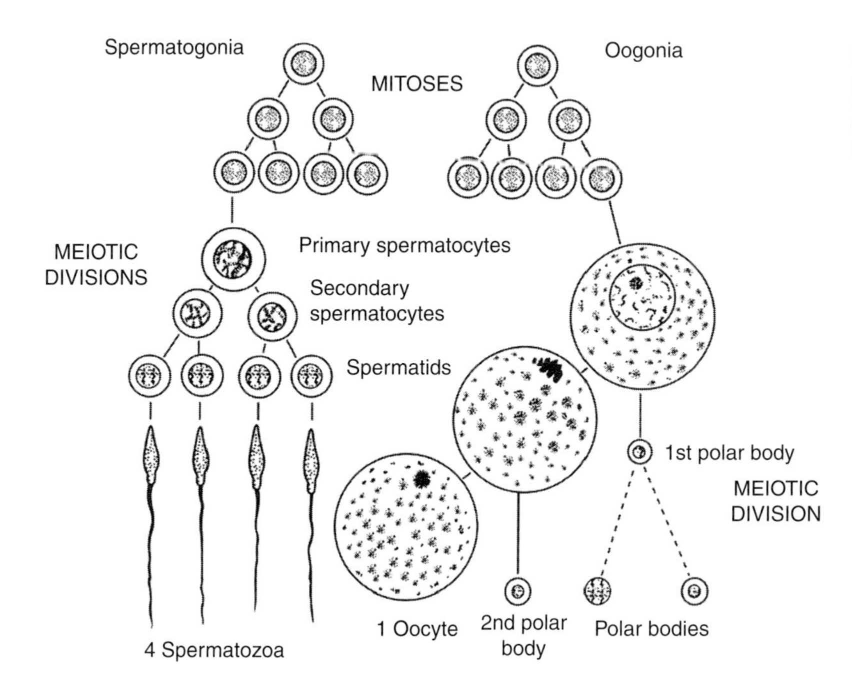

Figure 2.1 A schematic outline of gametogenesis in the male and female. Note that in the female, only one of the four cells produced at meiosis becomes a functional oocyte (from Dale 1983).

protein to fat ratio. In invertebrates and lower vertebrates, yolk is usually found in the form of small granules, evenly distributed throughout the cytoplasm and contributing around 20–30 per cent of the total oocyte volume. Amphibian yolk, by contrast, contributes up to 80 per cent of the oocyte volume and is organized into large flattened platelets. These platelets vary in size and are unequally distributed throughout the cytoplasm, the majority of the yolk lying at one pole, termed the vegetal pole. In teleosts, birds and reptiles, the yolk forms a compact central mass surrounded by a thin surface layer of cytoplasm with the nucleus in a thickened cytoplasmic cap at one end of the oocyte, termed the animal pole. Insect oocytes are similar but, in addition to the peripheral layer of cytoplasm, there is an internal mass of cytoplasm that contains the nucleus. In many animals, the material stored in the oocyte during growth appears to be synthesized in parts of the body distinct from the ovary and carried to the ovary in soluble form via the blood stream. For example, in vertebrates, proteins and phospholipids are produced in the liver. Once inside the oocyte, the Golgi apparatus processes these soluble yolk precursors into insoluble yolk granules.

Follicle Cells

During growth and maturation, oocytes are surrounded by a layer, or layers, of specialized somatic cells called follicle cells. The oocyte and its follicles are in close association (Figure 2.2). Gap junctions connect the follicle cells to the growing oocyte and serve as communicating devices allowing the passage of small molecules and ions between the cells. In later stages of growth, the surface of the oocyte is organized into microvilli, presumably to maintain a functional surface area to volume ratio. A dense fibrillar material appears between the oocyte and the follicle cells, which becomes the primary oocyte coat or vitelline membrane. However, the follicle cells remain in contact with the growing oocyte by way of long microvilli interdigitating with the microvilli of the oocyte. In mammalian oocytes, these microvilli create a radially striated layer called the zona radiata. In echinoderms and mammals, the microvilli are withdrawn shortly before ovulation, resulting in a continuous vitelline membrane; in others, such as bivalve molluscs, the vitelline coat remains perforated by the microvilli. The follicle cells serve to transfer materials and signals to the oocyte during growth and maturation.

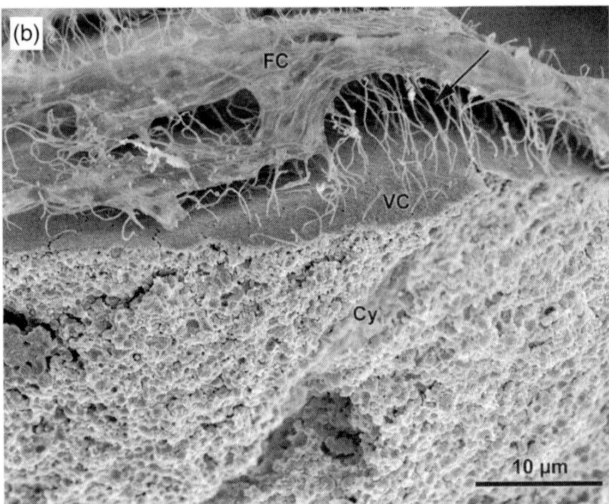

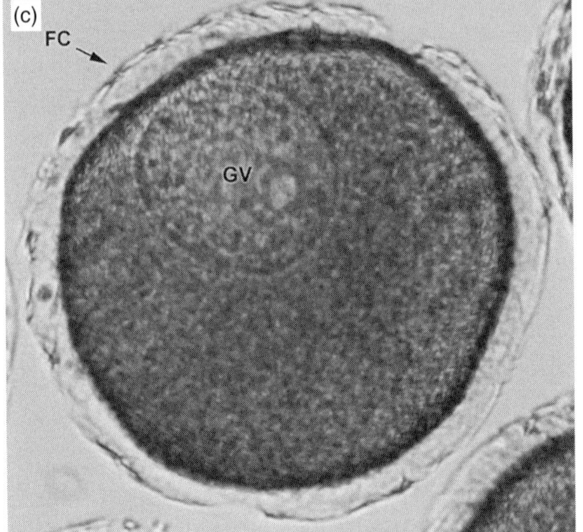

Figure 2.2 **(a)** An oocyte and its follicle cells in situ in the ovary of the starfish *Astropecten arranciacus*. N – Nucleus; FC – follicle cells; J – jelly layer; W – ovary wall. **(b)** A scanning electron micrograph of a fractured starfish oocyte to show the follicle cells (FC) and its projections to the oocyte plasma membrane. VC – vitelline coat. CY – cytoplasm. **(c)** A transmitted light photograph of a starfish oocyte showing the surrounding follicle cells (FC) and the oocytes large germinal vesicle (GV). By courtesy of Dr Luigia Santella, Stazione Zoologica. (A black-and-white version of this figure will appear in some formats. For the colour version, please refer to the plate section.)

In insects, molluscs and annelids, in addition to the follicle cells, we find nurse cells, which are concerned with the nutrition of the oocyte. Although the roles of follicle cells and nurse cells are similar, their origins differ. In the fruit fly *Drosophila,* each oogonium gives rise by four mitotic divisions to sixteen daughter cells, of which one grows to become the oocyte and the others become nurse cells. Thus, the nurse cells are sisters of the oocyte while the follicle cells are somatic in origin. The nurse cells are in continuity with the oocyte via cytoplasmic bridges, which in *Drosophila* are called fusosomes. The oocyte essentially incorporates the cytoplasm of the nurse cells and large amounts of RNA during its growth period (Figure 2.3).

Storing Information

In order to become fully competent for development, the oocyte must complete pre-maturation changes during its growth. This involves accumulation of specific macromolecules which are required later in the control of early embryogenesis. In the amphibian *Xenopus*, there is a considerable synthesis of rRNA during oogenesis, which falls off during maturation and is then undetectable until the beginning of gastrulation. This means that the ribosomes in the oocyte

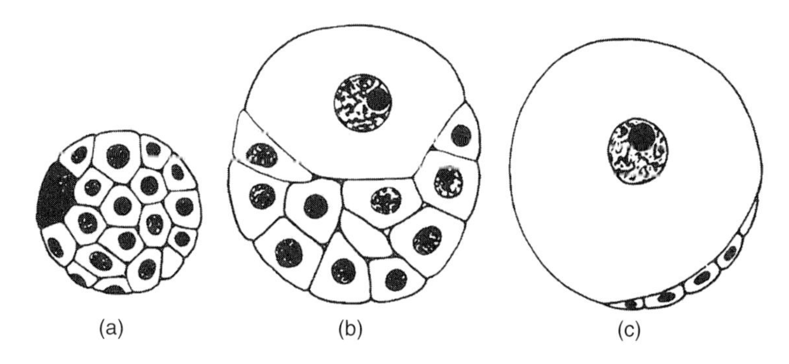

Figure 2.3 Three stages in the growth of the leech oocyte showing the incorporation of the cytoplasmic components of the nurse cells (from Dale 1983).

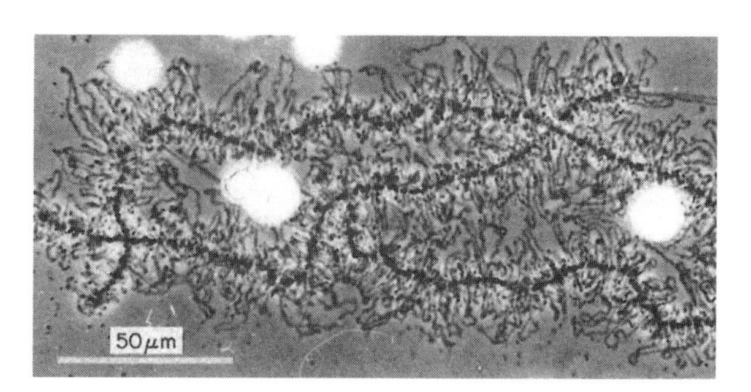

Figure 2.4 A lampbrush chromosome in an oocyte of the amphibian *Triturus* (from Dale 1983).

are present in sufficient quantity to support protein synthesis in an embryo containing many thousands of cells. How does the oocyte manage to synthesize such a huge amount of rRNA, corresponding to the total synthesis of some 200,000 liver cells? In amphibians, this is achieved by gene amplification: the rRNA genes are replicated many times over, forming several hundred copies. The germinal vesicle of the *Xenopus* oocyte contains many nucleoli; each nucleolus contains rRNA genes and is a site of rRNA synthesis. This mechanism of producing large amounts of rRNA in a relatively short period is by no means universally adopted, although gene amplification has also been detected in some invertebrate oocytes. In some insects, e.g. *Drosophila*, the nurse cells actively synthesize RNA, which is then transferred to the oocyte via the cytoplasmic connections. In the giant silk worm *Antheraea*, the DNA of the oocyte does not participate in RNA synthesis, and the nurse cells synthesize all of the RNA stored in the oocyte. This nurse cell-oocyte co-operation is certainly of great interest – but we should not forget that these two cell types are of common origin. Not all oocytes store large quantities of RNA. Mammalian oocytes contain a lesser amount of stored RNA, and new RNA is synthesized after activation of the zygotic genome during the early cleavage stages of the embryo. In the immature oocytes of some vertebrates and invertebrates, the diplotene chromosomes are extremely elongated with thin loops extending from the main axis – these are known as lampbrush chromosomes (Figure 2.4). The loops of lampbrush chromosomes are sites of intense RNA and protein synthesis and the RNA to DNA ratio is over 100 times the RNA to DNA ratio found in liver chromatin.

The Regional Organization of the Oocyte

Polarization indicates a differential distribution of morphological, biochemical, physiological and functional parameters in the cell. The appearance of polarization signals the triggering of the developmental programme. The growing oocyte does not have a homogeneous structure; in particular, many cytoplasmic organelles become segregated to various regions of the oocyte, and this regional organization determines some of the basic properties of the embryo. In all animal oocytes, the pole where the nuclear divisions occur, resulting in the formation of the polar bodies, is called the animal pole. The opposite pole, which often contains a high concentration of nutrient reserves, is called the vegetal pole. Many cytoplasmic inclusions and organelles are distributed according to this animal–vegetal (A–V) axis: yolk is usually more dense at the vegetal pole than at the animal pole, and in some animals, particularly amphibians, there is a density gradient of pigment granules. The familiar unpigmented region of the frog oocyte is, in fact, the vegetal hemisphere.

In cases where the heterogeneous organization of the cytoplasm cannot be detected, either by light or by electron microscopy, it can usually be inferred from developmental studies. During the 1930s, the Swedish zoologist Sven Horstadius carried out classical experiments on the sea urchin oocyte, in which one species of Mediterranean sea urchin, *Paracentrotus lividus*, was found to have a band of red pigment just below the equator towards the vegetal pole. This was the marker used to orientate the oocyte. When this oocyte was cut with a fine glass needle along the animal-vegetal axis (which is the plane of the first cleavage division), the two halves rounded up, forming two small cells. Both halves could be fertilized and potentially give rise to two normal plutei larvae. However, when the cut was equatorial, dividing the oocyte into animal and vegetal halves which were then fertilized, only the vegetal half gave rise to a complete pluteus. The animal half gave rise to a ciliated blastula-like sphere, which was incapable of gastrulation (Figure 2.5). The conclusion from these experiments was that in the unfertilized sea urchin oocyte there is an uneven distribution of factors essential for normal development along the A–V axis.

All oocytes have a polarized organization, and the embryo maintains this A–V axis throughout development. However, the axis is not always as rigidly determined as in the sea urchin. For example, the unfertilized ascidian oocyte may be cut into two halves along any plane and yet both halves, when fertilized, will give rise to normal larvae. This indicates that the fragments are capable of reorganizing themselves and developing a new A–V axis. In the molluscs *Dentalium* and *Lymnaea*, the area of contact between the oocyte and the ovary wall becomes the vegetal pole. Finally, in addition to A–V polarity, many oocytes express a bilateral symmetry either before fertilization, as in insects, or shortly afterwards, as in ascidian oocytes.

Oocyte Maturation

In the third and final phase of oogenesis, called maturation, the oocyte is prepared for ovulation and interaction with the spermatozoon. Maturation events are both nuclear and cytoplasmic. Throughout the growth period, the large oocyte nucleus, or germinal vesicle, is intact with the chromatin blocked at the first prophase of meiosis. The first sign of maturation is when the nuclear envelope breaks down, leading to the mixing of nucleoplasm with the cytoplasm and the migration of the semi-contracted chromosomes to the oocyte periphery, where they become arranged on the spindle. In most animals, meiosis is arrested a second time at the metaphase of the first or second meiotic division. The oocyte is now ready to be ovulated, and shortly after this, fertilization will occur. Thus in the majority of animals, the processes of oocyte meiotic maturation and fertilization overlap. Exceptions are some polychaetes,

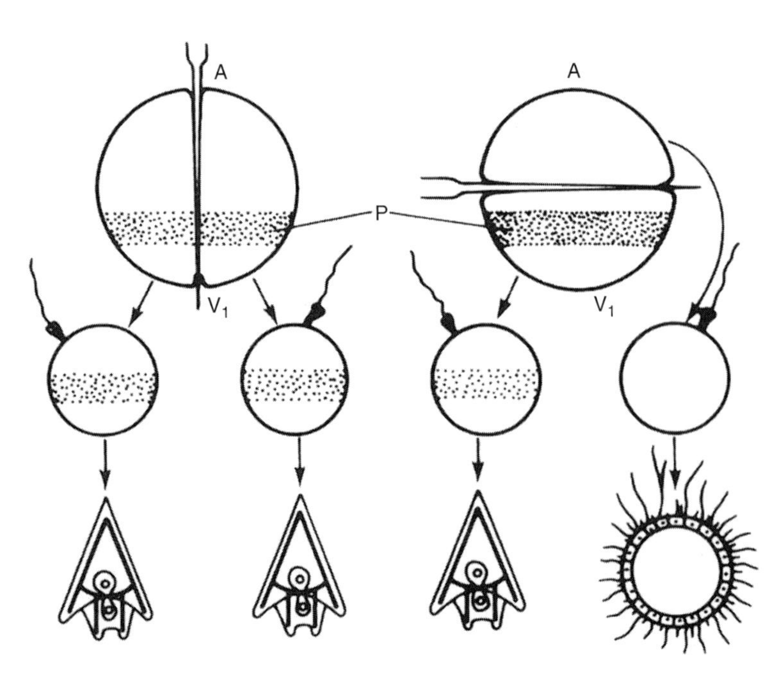

Figure 2.5 The experiments of Sven Horstadius, who demonstrated that polarized components of the vegetal hemisphere (V_1) are essential for development. An isolated animal half will only give rise to a ciliated blastula-like sphere (from Dale 1983).

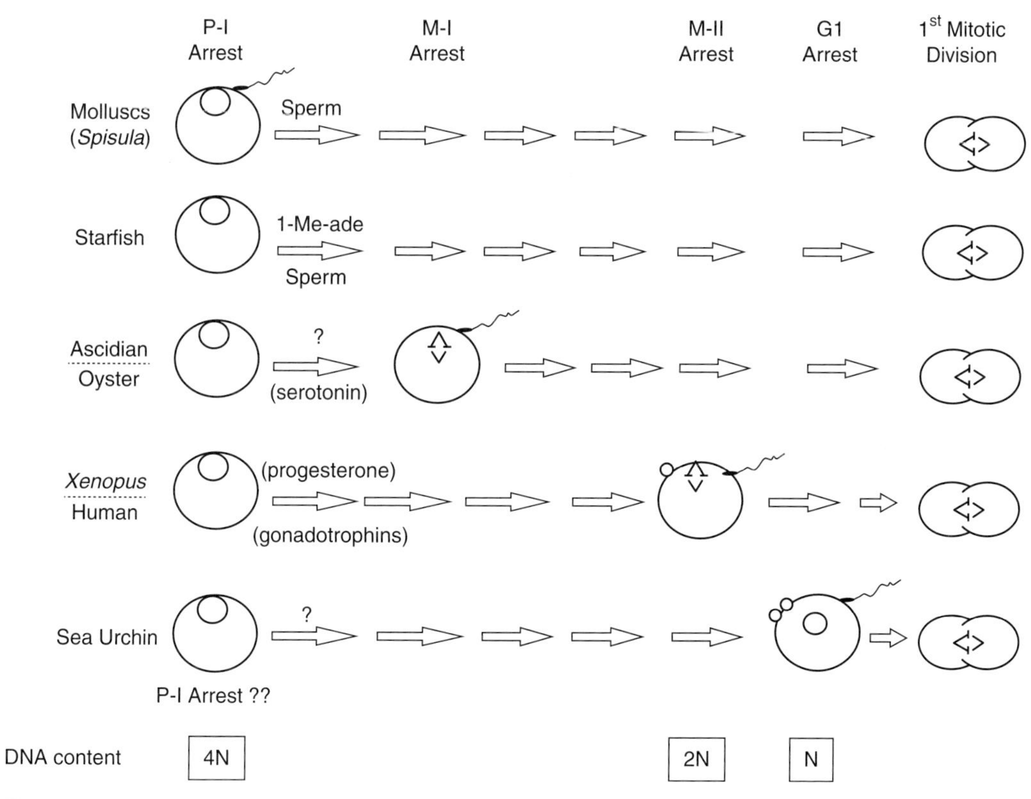

Figure 2.6 Meiotic arrest in the oocytes of various species. Normally the oocyte blocks twice in meiosis. First at prophase 1 and second at metaphase 1 or metaphase 2. Molluscs and sea urchins are exceptions.

where oocyte maturation is triggered by the fertilizing spermatozoon, and the sea urchins and coelenterates that have completed meiosis at fertilization (Figure 2.6).

Most of our information on oocyte maturation has come from studies in amphibians, starfish and mammals where hormones trigger the process. In amphibians, gonadotrophic hormones secreted by the pituitary glands control maturation and ovulation, while in starfish, the prime effector is a gonad-stimulating substance (GSS) produced by the nervous system that acts on the oocyte's follicle cells, stimulating them to release a second factor, which is 1-methyl adenine (1-MA). The maturation hormone 1-MA triggers meiosis re-initiation through a pathway involving receptors on the oocyte's plasma membrane.

The asymmetrical location of the germinal vesicle in the majority of animal oocytes marks the animal–vegetal axis (Figure 2.7). In the *Drosophila* oocyte, the nucleus repositions first to the posterior pole and then to the dorsal aspect regulated by cytoskeletal elements. In echinoderm oocytes, the centrosome appears to regulate the position of the GV. In most mammals, including humans, marmosets, cows and pigs, the GV is eccentrically placed, while in some rodents it often appears centrally positioned. Reports of centrally located GVs in mice and rats may be artefactual and due to the manipulation and isolation of the oocyte in vitro. However, even in these cases of centrally localized GVs, the nucleus seems to be anchored in place to the animal pole by a microtubule organizing centre (MOTC). Mouse oocytes have a highly dynamic actin network that mediates long-range transport of vesicles that are positive for the small GTPase RAB11. These vesicles move up to 30 μm along the actin filaments to the oocyte surface and function as cytoskeletal modulators. Rab11, myosin Vb and the actin network are required for the asymmetric positioning of the spindle during mouse oocyte maturation.

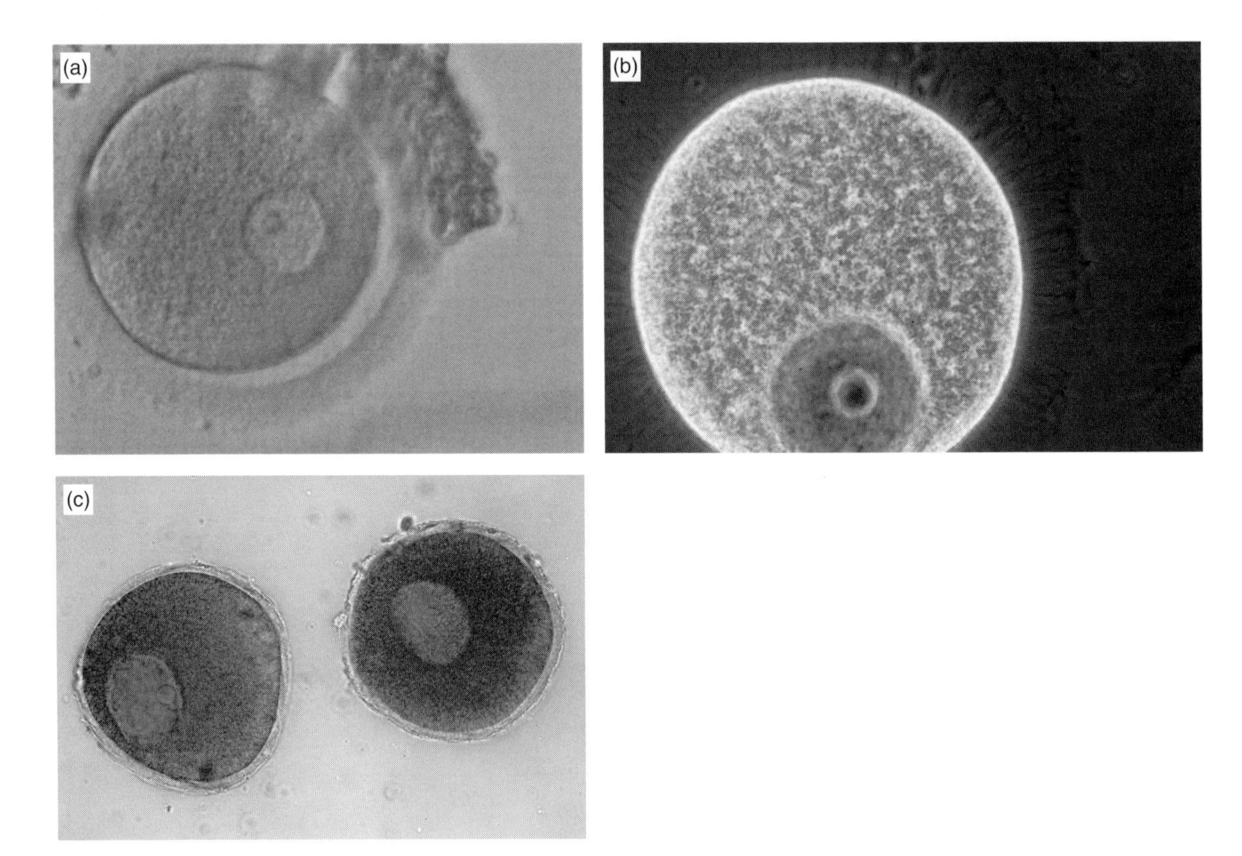

Figure 2.7 The eccentrically located germinal vesicle in Prophase 1 oocytes of the **(a)** human, **(b)** sea urchin and **(c)** starfish. The latter is courtesy of Dr Keiichiro Kyozuka. (A black-and-white version of this figure will appear in some formats. For the colour version, please refer to the plate section.)

Spermatogenesis

Spermatogenesis is composed of three phases: proliferation of cells by mitosis, reduction of the chromatin content by meiosis and finally differentiation of the spermatozoon in a process called spermiogenesis. In the male, and in contrast to the female, four spermatozoa are produced from one primary spermatocyte (see Figure 2.1). Differentiation of the spermatid into a spermatozoon is remarkably similar across the whole metazoa (Figure 2.8). Small vesicles in the Golgi develop, each one containing a granule. These enlarge and coalesce, forming the acrosomal vesicle that attaches itself to the nucleus. In some animals the acrosomal vesicle remains attached to the head of the nucleus; in others it spreads out around the nucleus, forming a cap. At the same time the two centrioles move to the plasma membrane of the spermatid and one, the distal centriole, becomes attached to the plasma membrane; from here the tail is generated. The centrioles then locate to the nucleus, assuming a position opposite the acrosomal vesicle. Mitochondria form the midpiece, while the remaining cytoplasm forms a droplet and is discarded.

Ion Channels and Gametogenesis

Gametes possess a wide variety of ion channels and transporters which are essential for their function. Modifying or eliminating – partially or totally – any of these plasma membrane elements would of course be detrimental to development. Echinoderms have been used extensively to study the electrophysiological properties of the oocyte plasma membrane. In the germinal vesicle stage of starfish oocytes, three types of voltage-gated currents have been identified: an inward Ca^{2+} current, a fast transient K^{+} current and an inwardly rectifying K^{+} current. During hormone-induced in vitro maturation with 1-MA, the Ca^{2+} current increases; whereas both K^{+} currents decrease in amplitude, leading to a decrease in membrane conductance and a

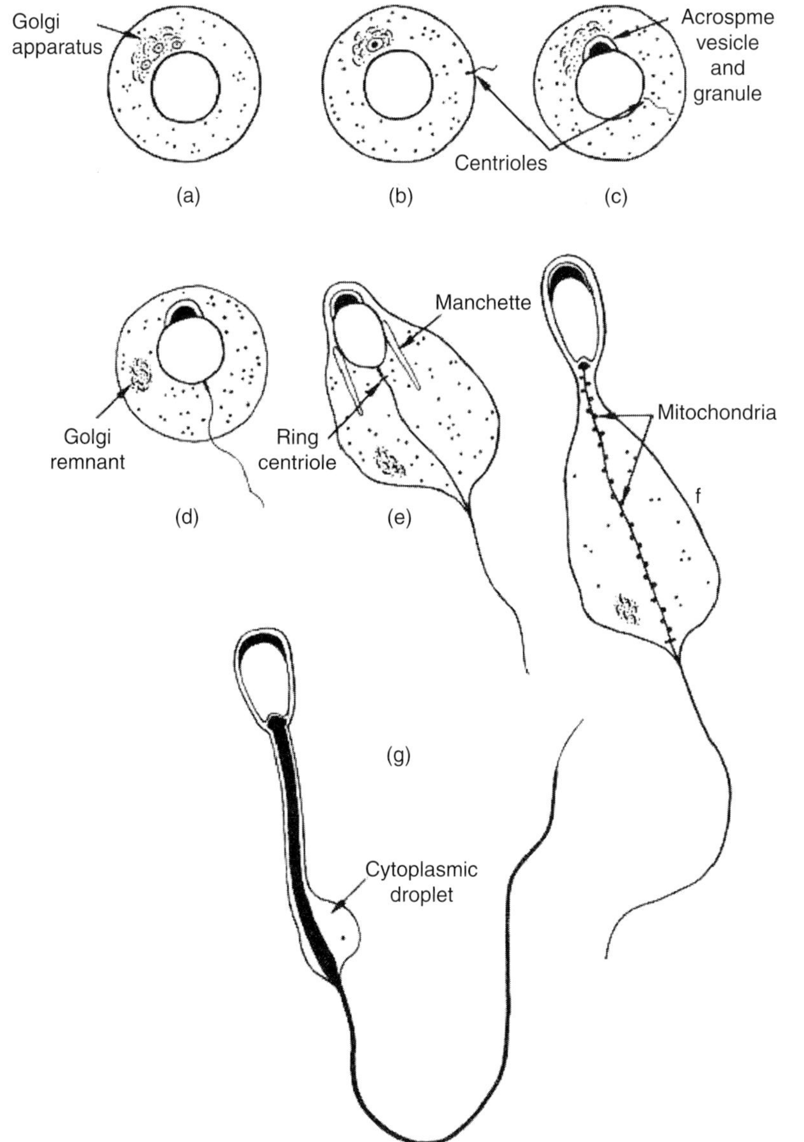

Figure 2.8 Spermiogenesis in the rabbit. Small vesicles in the Golgi form the acrosomal vesicle, while the flagellum develops from a centriole (from Austin 1965).

depolarization of the resting potential. A transient inward Na^+ current, a transient inward Ca^{2+} current and an inwardly rectifying K^+ current have also been identified in ascidian oocytes. At germinal vesicle breakdown (GVBD) in ascidians, the Na^+ current diminishes, while the Ca^{2+} current reaches its peak. Since the calcium current is composed mainly of L-Type calcium channels, it may play a role in regulating cytosolic calcium at maturation.

In amphibians, K^+ and Cl^- voltage-gated currents present in immature oocytes decrease in mature oocytes and are replaced by a Na^+ current, while a voltage-dependent hydrogen current has been described in axolotl oocytes. Preventing external calcium from entering immature oocytes reduces the efficiency of both GVBD and the release of calcium after activation, suggesting that calcium currents during prophase 1 potentiate GVBD and fill the internal Ca^{2+} stores of the oocyte in preparation for fertilization. In *Xenopus* and *Caenorhabditis elegans*, a chloride current that increases the efficiency of GVBD has been identified.

In mammalian oocytes, there are also changes in ion permeability of the plasma membrane which appear to be due to L-type Ca^{2+} channels again suggesting a role for external sources of Ca^{2+} in the mobilization of intracellular stores. In mature human oocytes, the patch-clamp technique shows that the most frequent channel is a 60 pS non-inactivating, K^+-selective pore, which is activated by depolarization.

Despite their small size, techniques such as patch and voltage clamp, ion-sensitive fluorescent indicators, immunocytochemistry, pharmacology and DNA recombinant technology have demonstrated a role for ion channels in sperm function. Single-channel recording techniques allowed the identification of K^+ and Cl^- channel activity in the sea urchin sperm plasma membrane, and this was extended to several animals, revealing the presence of K^+, Ca^{2+} and Cl^- channels. K^+ and Ca^{2+} currents have been measured by the whole-cell clamp technique in rat spermatogenetic cells, and the negative resting potential of rat spermatids was shown to be due to Cl^- and K^+ conductance with a minor contribution from Na^+ conductance. Cl^- conductance regulates spermatogenesis in the roundworm *Caenorhabditis elegans*, while in mouse spermatogenetic cells, a pH dependent Ca^{2+} permeability factor and a series of K^+-selective currents are correlated with the function of mature sperm. T-type Ca^{2+} channel expression is regulated during mouse spermatogenesis. Voltage-gated sodium channels appear to be required for motility in spermatozoa, whereas voltage-gated potassium and anion channels are associated with capacitation. Calcium channels are also required for motility and the calcium influx that primes hyperactivation and the acrosome reaction necessary for fertilization. Spermatozoa appear to have two types of calcium channels. The first, a sperm-specific channel called CatSper located on the flagellum of human spermatozoa is involved in hyperactivated motility; whereas L-type voltage-gated calcium channels located on the acrosome control the rapid influx of calcium required for the acrosome reaction. Less is known about ligand-gated channels in spermatozoa; however, an ATP-gated sodium channel has been defined, although its function is not known.

Gametogenesis in Mammals

Primordial Germ Cells

In mammals, the early gonad is similar in males and females and is derived from somatic mesenchymal tissue precursors called the genital ridge primordia. In both sexes, the primordial germ cells originate outside the gonad. In the human embryo, primordial germ cells can be seen at three weeks in the yolk sac near the base of the developing allantois (Figure 2.9). These then increase in number by mitosis and migrate to the connective tissue of the hind gut and on into the gut mesentery. Within 30 days of conception, the majority of cells have passed into the region of the developing kidney and then into the adjacent genital ridge primordia. The migration of PGCs is completed by six weeks and occurs by amoeboid movement. PGCs may locate the genital ridge following a chemotactic gradient. At about the same time, a second group

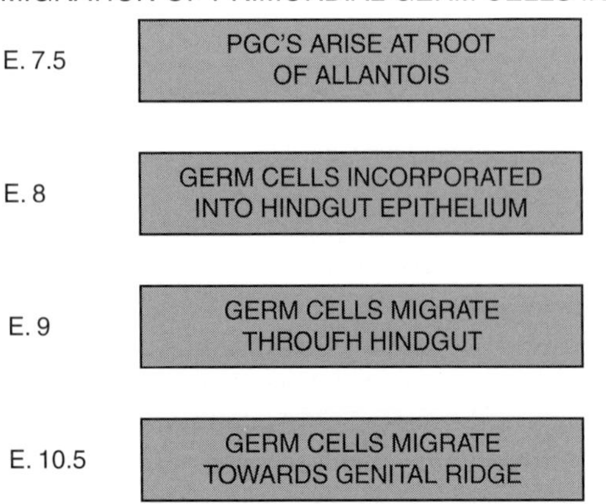

Figure 2.9 Migration of primordial germ cells in the mouse. The embryonic day is indicated to the left.

of cells starts to migrate in columns into the genital ridges called the primitive sex cords. In males, the Sry gene is expressed in these cord cells, and they become the testis cords and eventually the Sertoli cells, while the PGCs will give rise to spermatozoa.

In females, the Sry gene is not expressed in the sex cords, and they become the granulosa cells of the primordial follicle; whereas here the PGCs become the oogonia. The third wave of migratory cells comes from the mesonephric primordia and forms several testis tissues in the male, such as vascular tissue, the Leydig cells and the seminiferous tubules. In the female, myoid cells do not migrate; however, there may be comparable cells that develop into the thecal cells around the developing follicles. In mammals, the initial decision to become an ovary or testis depends on the expression of the Sry gene, while full development of the gonad requires the presence of a normal complement of the sex chromosomes, i.e. two X chromosomes in the female and not more than one X chromosome in the male.

Spermatogenesis in Mammals

At puberty, interphase germ cells start to proliferate by mitoses and at this stage are called spermatogonial stem cells. Some of these periodically emerge to go through a series of mitoses: these are the A1 spermatogonia. These form a clone of daughter cells, the size of which depends on the species. For example, in the rat there are six divisions leading to 64 cells. After the fifth division in the rat, type B cells are produced that become the resting primary spermatocytes (Figure 2.10). All the spermatocytes derived from one A1 spermatogonium are linked by cytoplasmic bridges which remain through meiosis until the mature spermatozoa are released. This first phase of spermatogenesis takes place in the basal intratubular compartment of the testis (Figure 2.11).

After the final mitotic division, the primary spermatocytes duplicate their DNA content and move into the adluminal compartment to enter into meiosis. In this compartment, they undergo two meiotic divisions to form, first, two daughter secondary spermatocytes, and eventually four early round spermatids. Thus, in the rat, from the 64 primary spermatocytes entering meiosis, a maximum of 256 spermatids are formed. Many are lost by defective mitotic or meiotic divisions. All the spermatids in a cluster are connected by cytoplasmic bridges and are now haploid. The round spermatids now undergo a morphological transformation from a small round cell into an elongated structure with a condensed nucleus and a flagellum, with a specific shape that

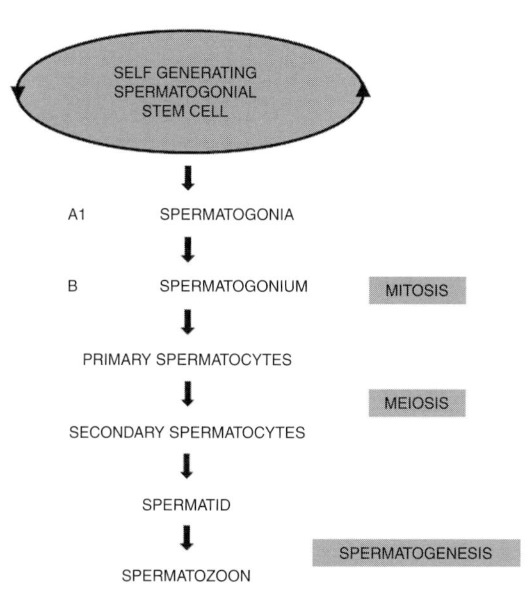

Figure 2.10 Spermatogenesis in the rat.

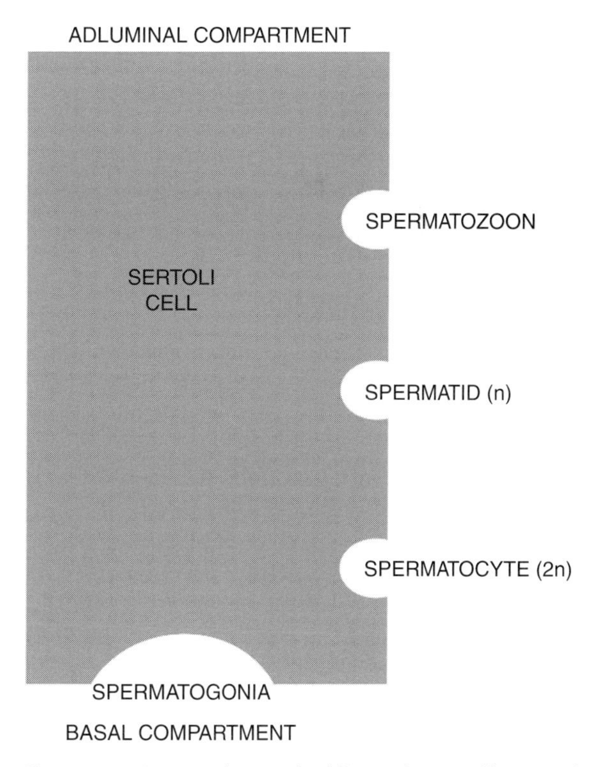

Figure 2.11 Diagram showing the different phases and location of spermatogenesis in the mammal. The Sertoli cell regulates modelling and maturation.

varies from species to species. Cytoplasmic reorganization gives rise to the tail and the midpiece containing the mitochondria and associated control mechanisms necessary for motility, the acrosome is constructed from Golgi membranes, and a residual body casts off excess cytoplasm. The centrioles of the spermatid are reduced to a central core, and they discard all of the pericentriolar material. Sperm modelling is probably regulated by the Sertoli cells, and the cells are moved to the centre of the tubular lumen as spermatogenesis proceeds. In humans, spermatogenesis is complete in 64 days. The rate of progression of cells through spermatogenesis is constant and unaffected by external factors such as hormones. mRNA production and translation is active throughout spermatogonial mitosis and meiosis except on the sex chromosomes. Autosomal transcription stops when the cell reaches the elongated spermatid phase. The cessation of transcription is due to the condensation of the chromatin following the replacement of nuclear histones with protamines.

The process of spermatogenesis in mammals is controlled by the hypothalamic gonadotropin-releasing hormone (GnRH) which stimulates the secretion of follicle-stimulating hormone (FSH) and luteinizing hormone (LH). FSH binds to receptors on the Sertoli cells and is required for the synthesis of inhibin and activin, and the stimulation of androgen receptors. In fact, FSH and androgens act synergistically on the Sertoli cells to support its function in spermatogenesis, while LH stimulates the Leydig cells to produce testosterone.

Oogenesis in Mammals

We may divide oogenesis in mammals into three phases; the pre-antral phase (primary), the antral follicle phase (secondary or Graafian) and finally the pre-ovulatory follicle.

The ovary in the four-month-old human foetus contains around seven million oogonia, which gradually declines to about one million at birth and 300,000 when the girl reaches menarche.

At three months gestation, the oocytes start meiosis and progress to diplotene of first meiosis and form the resting dictyate primary oocytes. At this point each oocyte is surrounded by a single layer of somatic cells, the pre-granulosa cells. These oocytes within their surrounding follicles are the primordial follicles, which will remain for the woman for up to 50 years, i.e. her reproductive lifespan. At puberty, recruitment of some of these primordial follicles begins. The recruited follicle grows from a diameter of 20 μm to several hundred μm, and the oocyte itself grows in diameter from 10 μm to about 100 μm. The growth phase essentially involves synthesis and storage of large amounts of proteins, metabolic substrates and polyadenylated mRNA. The human oocyte accumulates 1,500 pg of total RNA and the number of mitochondria increases to 100,000, while the number of ribosomes increases to 10^8.

During this growth period, the surrounding granulosa cells divide mitotically, and the zona pellucida, a glycoprotein coat synthesized by the oocyte, is secreted between the oocyte and the surrounding somatic cells. Gap junctions connect the somatic cells and the oocyte allowing the transfer of amino acids, nucleotides, rRNA and mRNA into the growing oocyte. The somatic cells are also connected with each other, forming a large network. This network is important for nutrition since at this stage the granulosa does not contain blood vessels. The pre-antral follicle also increases in size due to the formation of the theca, an outer layer of cells attached to the membrane propria.

The production of primordial follicles is controlled by retinoic acid (Figure 2.12) and is inhibited by both oestrogen and progesterone. The primordial follicle pool depends on the dynamics of atresia, activation and maintenance by repressive signals such as anti-mullerian hormone, phosphatase and tensin homolog (PTEN) and Foxo3. Stem cell factor, also known as KIT-Ligand, induces the transition from the primordial follicle to the primary follicle through a phosphatidyl-inositol-3-kinase (PI3 K) pathway. Other factors that appear to induce primordial follicles are basic fibroblast growth factor (bFGF) and transforming growth factor β (TGFβ). Anti-mullerian hormone, a member of the TGFβ super family has the capacity to block antral follicles. Communication between the oocyte and its surrounding somatic cells is in both directions. In fact, the oocyte secretes soluble growth factors that control the destiny of the granulosa cells and the cumulus cells.

The gonadotrophins FSH and LH are essential for antral development and act through cytokines and sex steroids. At the beginning of this phase, the cells of the theca interna bind LH and only the granulosa cells bind FSH, whereas towards the end of this phase, and an essential step for entry into the pre-ovulatory phase, the outer layers of the granulosa cells bind LH. The theca

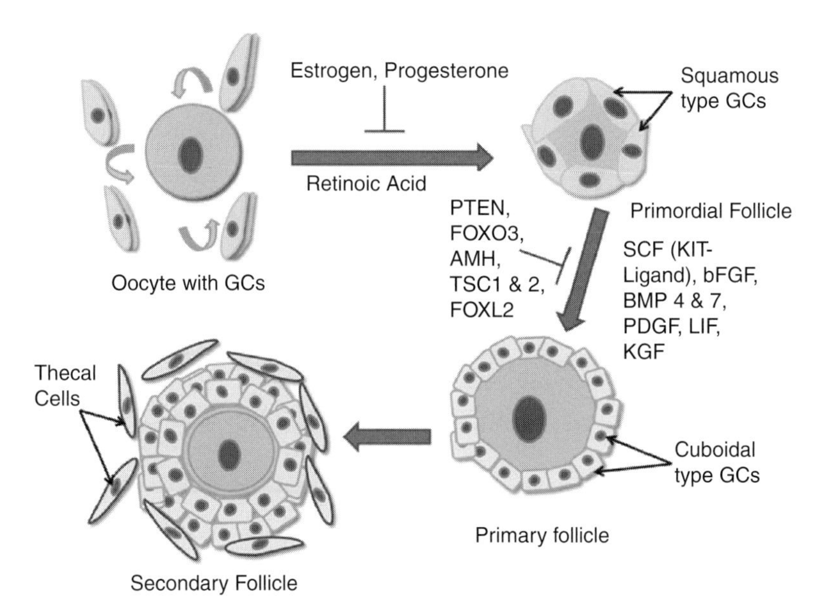

Figure 2.12 Progression from the primordial follicle stage to the secondary follicle in mammals (from Huang and Wells 2011).

cells produce androgens, which serve as a substrate for conversion to oestrogens and, in conjunction with FSH, stimulate the growth of the granulosa and therefore follicular growth. The gonadotrophins also stimulate the production of cytokines such as IGF, inhibins and SCF that support follicular progression and others such as leptin, IGFBP's MIH and TNFα that depress follicular development.

Growth of the secondary follicle (antral) depends mainly on the effect of FSH, and the oocyte at this stage continues to synthesis RNA and store proteins. The granulosa cells secrete muco-polysaccharides that together with serum transudate forms the follicular fluid and eventually the follicular antrum. Although follicular fluid is formed in part from blood serum, the composition of follicular fluid differs considerably from plasma, with lower glucose and lipid concentrations, different amino acid concentrations, and the presence of steroid binding proteins. Follicular fluid has been shown to be an excellent buffer of pH, more so than blood serum or artificial culture media. Follicular fluid also contains buffers of free radicals, and many transport proteins, signals and hormones. In fact, pre-incubation of human oocytes in follicular fluid for an extended period after oocyte retrieval appears to increase the potential of these to form embryos and implant, suggesting that the composition of artificial culture media still does not precisely mimic the physiological environment.

The antral cavity reaches a diameter of 10–12 mm over a period of eight to ten days and eventually the pre-ovulatory follicle reaches 25 mm. The oocyte at this stage is blocked in meiosis by cyclic adenosine monophosphate (cAMP) and a cAMP-dependent protein kinase A, which regulate the activity of MPF. A surge in LH stimulates the somatic cells in the follicle promoting a decrease in cAMP in the oocyte and inducing meiotic resumption. The oocyte that is now held in the fluid-filled follicle by a thin stalk of cumulus cells extrudes the first polar body and arrests again at metaphase 2. At ovulation, the oocyte is transcriptionally silent and remains more or less so until the four-to-eight-cell stage of embryonic growth. The follicle ruptures, owing to the activity of metalloproteineases, releasing the oocyte, while the follicle over a 13-day period luteinizes to form the corpus luteum.

Only one in a thousand primordial follicles complete oogenesis to reach the stage of a corpus luteum. Most become atretic and die in a process that is not fully understood. In the human, at the beginning of each menstrual cycle, up to 20 antral follicles are recruited for development, and this is regulated by FSH; however, usually only one becomes the dominant follicle. In response to the LH surge, the cumulus oophorus expands almost 40-fold, due to the accumulation of a voluminous extracellular matrix. Simultaneously, the gap junctions between cumulus cells, and between cumulus and granulosa cells, disappear, with the deposition of a large, mucified extracellular matrix. This extracellular matrix results from massive de novo synthesis of hyaluronic acid by the cumulus cells after the LH surge. Cumulus cells express the ganglioside GM3, growth-factor receptors, prostaglandins and vascular endothelial growth factor (VEGF).

Gametogenesis in Lower Vertebrates

In the young frog, primary spermatogonia multiply by mitosis to form a pool of juvenile primary spermatogonia which transform into spermatogenic stem cells. Active spermatogenesis only occurs in the adult, where the primary spermatogonium, a large cell of some 15 μm enclosed by one to three Sertoli cells, will divide by mitosis to form two kinds of primary spermatogonia: pale spermatogonia, which are the source of the next generations of primary spermatozoa, i.e. they retain their ability to divide mitotically; and the dark primary spermatogonia that will give rise to the secondary spermatogonia. At this stage in *Xenopus* the transient pre-meiotic rRNA amplification takes place. A primary spermatogonium generates a clone of secondary spermatogonia which are connected by cytoplasmic bridges and enclosed in a cyst formed by Sertoli cells. A cyst may contain hundreds of cells, and after the fifth mitotic division, the cells have two choices, continue dividing mitotically or enter into meiosis. The secondary spermatogonia now transform into primary spermatocytes and undergo the first meiotic division to become secondary spermatocytes, followed by the second meiotic division to become spermatids. The early spermatids are still connected by cytoplasmic bridges and are enclosed in a cyst until they start to elongate, a flagellum is formed and the cysts open (Figure 2.13). In the differentiating sperm cell, one centriole, the proximal centriole with its pericentriolar material, is situated close to the nucleus, while the distal centriole is perpendicular to the long axis of the spermatid. Spermatogenesis in anurans is also regulated by the hypothalamo-pituitary-gonadal axis as shown in Figure 2.14.

Spermatogenesis in urodeles is similar to that in anurans and the reduction in size of the germs cells during this process quite marked. Here the primary spermatogonia are 40–50 μm in diameter, reducing to 35–45 μm as secondary spermatogonia, 18–20 μm as secondary spermatocytes and 14–17 μm as early spermatids.

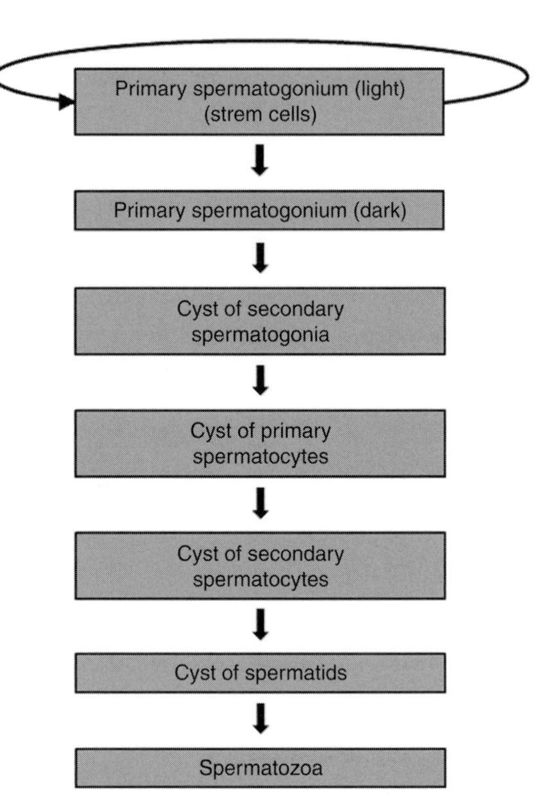

Figure 2.13 Spermatogenesis in amphibians. Light primary spermatogonia renew the pool of stem cell spermatogonia, while dark spermatogonia progress to form the clones of secondary spermatogonia followed by meiosis leading to mature spermatozoa.

Frog ovaries may contain thousands of follicles, each containing a diplotene oocyte. Primary oogonia in frog ovaries act as stem cells, continually renewing the pool of germ cells before each breeding season. Over a period of two to three years, the oocytes grow by accumulating RNAs and yolk until ovulation. The primary oogonia are 15–20 μm in diameter; they are surrounded by a few flat somatic cells and undergo several mitotic divisions. Some primary oogonia, become precursor cells for secondary oogonia and undergo synchronous mitotic divisions forming a cluster of cells connected by cytoplasmic bridges and surrounded by follicle cells. These oogonia nests are analogous to the cysts of secondary spermatogonia in the testis. The intercellular bridges between secondary oogonia are up to 1 μm in diameter and contain both actin and spectrin in their walls. In the next stage the oocytes are called primary oocytes and are at various stages of first prophase.

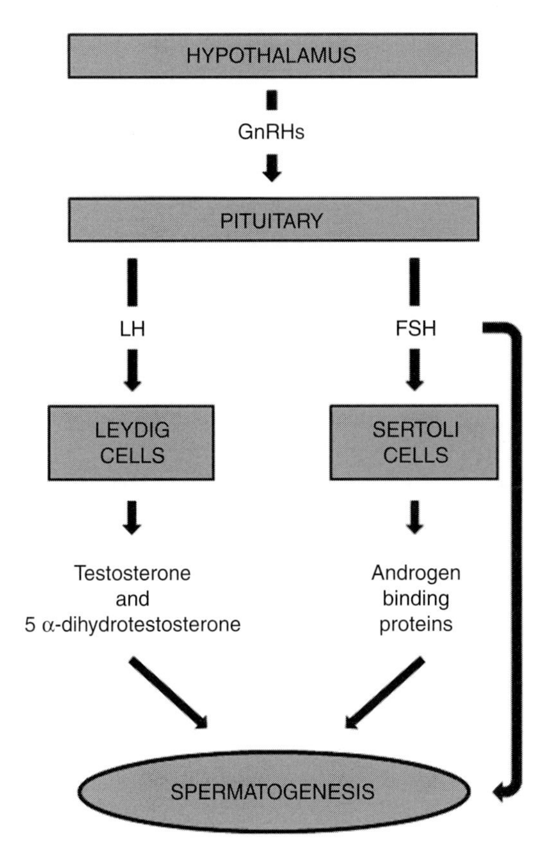

Figure 2.14 The hypothalamo-pituitary-gonadal axis regulating spermatogenesis in amphibians.

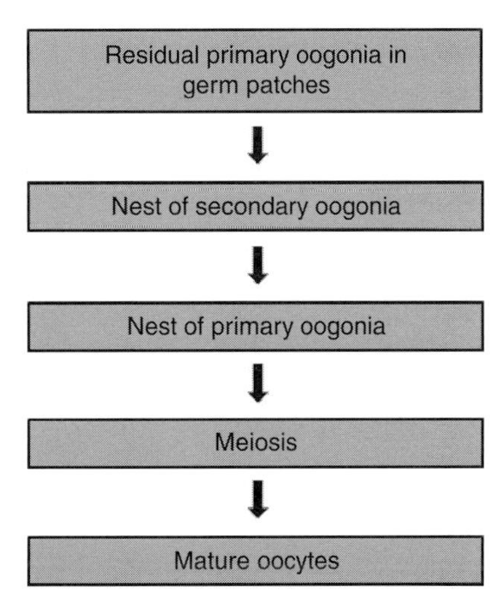

Figure 2.15 Oogenesis in amphibians. Oogonia form nests of primary oocytes where the cells are connected by cytoplasmic bridges. Amphibian oocytes are ovulated at metaphase 2.

The chromosomes start to condense, and, at the end of pachytene, the cytoplasmic bridges disappear and each germ cell becomes surrounded by proliferating follicle cells to form its own follicle. This stage, in which the oocytes are in diplotene, constitutes the longest phase in oogenesis and can last from a few months to several years depending on the species. Diplotene oocytes are the fastest growing stage in oogenesis, increasing in size enormously, and this growth phase may be divided into six stages. In stage 1 *Xenopus* oocytes, the cytoplasm is transparent and rich in ribosomes, with the cell measuring 50–300 μm, growing to 300–450 μm at the opaque stage 2 phase where the oocytes contain cortical granules and mitochondria. The stage 3 oocytes are 450–600 μm, and here the first signs of yolk deposition and lampbrush chromosomes are seen. The cell reaches 600–1000 μm at stage 4, where gradients in intracellular components, such as pigment granules and yolk platelets can be noted, while at stage 5 the cell diameter reaches 1000–1200 μm and the intracellular polarized distribution of cytoplasmic components is quite marked. At this stage the nucleus moves towards the animal pole of the cell. Stage 6 oocytes are fully grown, measuring 1200–1300 μm (Figure 2.15).

During growth of the oocyte cytoplasm the nucleus also grows keeping the cytoplasmic/nuclear ratio constant. The chromatin of the diplotene oocyte is highly active in transcription. Lampbrush chromosomes, first seen as early as stage 1, are two homologues joined by chiasmata, with each chromosome made up of two parallel sister chromatids. Where transcription is active, the chromatin fibres are loosely arranged loops (Figure 2.4). This high activity is necessary to provide the large quantity of RNA to be stored in the growing oocyte to support early development until the mid-blastula transition when the translation of the zygotic genes is in full swing. The lampbrush chromosomes produce mRNA that is stored throughout the cytoplasm as poly(A)RNAs. In fully grown *Xenopus* oocytes, only 2 per cent of ribosomes are found in polysomes; the rest are activated during early development. Amphibian oocytes produce 300,000 ribosomes per second, which is tens of thousands more than a typical somatic cell. Oogenesis in amphibian is regulated by the hypothalamo-pituitary-gonadal axis with the same hormones as those found in males and other vertebrates.

Gametogenesis in Invertebrates

Compared to the vertebrates, our knowledge of gametogenesis in invertebrates is quite fragmentary, probably due to the great diversity found in these phyla that represent 95 per cent of all animals.

Annelids

Annelids are commonly found in all habitats, and although the majority of polychaetes are gonochoric while the oligochaetes are hermaphroditic, both possess well-defined permanent ovaries. Oogenesis is either extraovarian, where small previtellogenic oocytes are released into the fluid-filled coelom to complete vitellogenesis, or intraovarian, where vitellogenesis is completed in the ovary. In the late previtellogenic stage, the oocyte surface develops long microvilli, the Golgi complex proliferates and the rough endoplasmic reticulum (RER) is prominent in parallel and whirled arrays. The vitellogenic phase occurs during the diplotene stage of first meiotic prophase in which the oocyte increases in volume due to the storage of glycogen, lipid droplets and membrane-bounded yolk platelets. Yolk is produced both autosynthetically by the Golgi and RER using low molecular weight oocyte precursors produced outside the oocyte and heterosynthetically by the incorporation of extraovarian high molecular weight yolk proteins. Polychaete intraovarian oocytes are in contact with somatic follicle cells until ovulation, and these are thought to contribute to biosynthetic processes.

Most annelids lack a permanent testis; spermatogonia or spermatocytes are released into the coelom from germ cells lining the peritoneum. This simple 'testis' contains only spermatogonia and stem cells, although in some annelids, such as lumbricids, later stages of spermatogenesis may be found. In the Echiura, there is no fixed testis and the germ cells float freely in the coelom. As a generalization we may say that in annelids maturation of spermatozoa takes place in the coelomic fluid. The sperm develop in syncytial masses and the number of spermatids in these clusters may range from a few to several hundreds.

Sea Urchins

In the sea urchin gonads of both sexes, nutritive phagocytes (NPs) play a similar role to Sertoli cells in vertebrates (Figure 2.16). They provide a structural and nutritional environment for the germinal cells. In the ovary, each NP encloses a single growing vitellogenic oocyte; in the testis, NPs cooperate to form large basal incubation chambers to supply nutrients to spermatogenic cells at various stages of spermatogenesis. Again, we see the importance of the interaction of somatic and germinal cells in the reproductive process. In the male sea urchin, only four interconnected spermatids result from spermatogenesis, which contrasts to the mammals where hundreds may be interconnected. As spermatogenesis proceeds the fully differentiated spermatozoa are stored in the lumen of the testis. The full process of spermatogenesis in sea urchins takes about 12 days. At the end of gametogenesis, the NPs have exhausted their nutrients and are reduced in size while the gonadal lumen is full of stored differentiated gametes. During gametogenesis, several proteins are stored in the NP, with the principal protein being a major component of yolk granules called major yolk protein (MYP). In the sea urchin, both males and females produce MYP, which is synthesized in the digestive tract and in the NPs and acts as an amino acid reserve.

Both primary oocytes and spermatocytes undergo meiosis while in the NP incubation chambers. Some of the genes in the sea urchin genome involved in oogenesis and meiosis are shown in Figure 2.17.

Insects

In *Drosophila*, the germ cells divide in the germarium leading to the formation of 15 nurse cells and the oocyte surrounded by a single layer of follicle cells (Figure 2.18). As the oocyte enlarges, the follicle cells move posteriorly making a thick columnar layer of cells over the oocyte. A group of follicle cells at the anterior region of the oocyte will form the micropyle. The growing oocyte receives yolk from the follicle cells and other material from the nurse cells. Some follicle cells then surround the oocyte totally and secrete the vitelline membrane and the chorionic layers of the eggshell. There are constantly changing patterns of differential and spatial gene expression in these migrating follicle cells, including those for transcription factors, signalling molecules and effector molecules. Several genes are known to be important in the establishment of polarity in the *Drosophila* oocyte. The terminal follicle cells at both ends of the developing oocyte chamber are established via Notch signalling. Another gene, Fringe, is important for the encapsulation and separation of individual cysts in

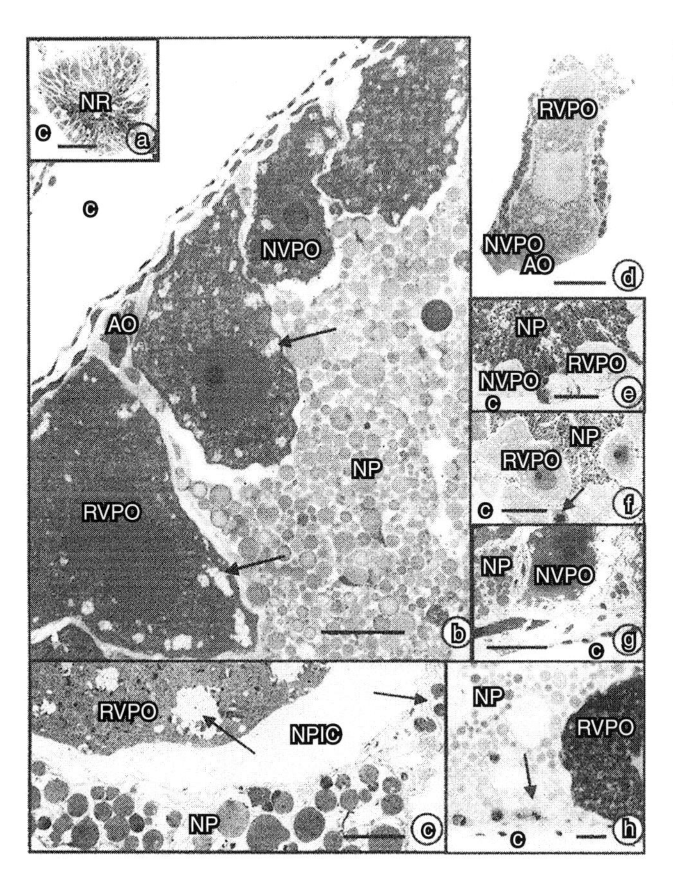

Figure 2.16 Nutritive phagocytes in sea urchin gonads have a role similar to Sertoli cells in mammals (from Walker et al 2005).

the germarium and for maintaining the organization of the follicle cell layer. Signalling by the epidermal growth factor receptor (EGFR) determines the fate of terminal follicle cells by inducing interactions between the germline cells and the surrounding follicle cells.

Finally, in *Drosophila*, many stored RNA molecules are not translated until after fertilization; in addition, they are localized to specific areas of the oocyte and are responsible for both the establishment of the anterior-posterior axis and germ cell determination. These include *bicoid*, which is necessary for anterior determination, *nanos* for posterior abdomen determination and *oskar* for posterior abdomen and germ cell determination. They are held in place and repressed due to the binding of proteins, such as Staufen, to their 3' untranslated regions.

In the male insect, a pair of testis, a pair of lateral ducts and a median duct open into the eighth segment via a penis. In each testis are sperm tubes containing the gametes at various stages of development. In many insects, spermatozoa are transferred to the female in spermatophores which are deposited directly into the female tract, while in spiders it is deposited on the ground and taken up by the female. The sperm are stored in the spermatheca and remain there until ovulation.

Nematodes

Most nematodes are dioecious and the testis is a single tubular organ formed of a blind end where the germ cells are formed and an elongated region where the spermatocytes differentiate. The spermatozoa, unlike most animals, lack flagella and assume various shapes from round to elongated. They are immotile when stored and become active with amoeboid movement within the female. Boveri in 1909 showed that the somatic and germ cell lines are established early in development, with the somatic cells undergoing

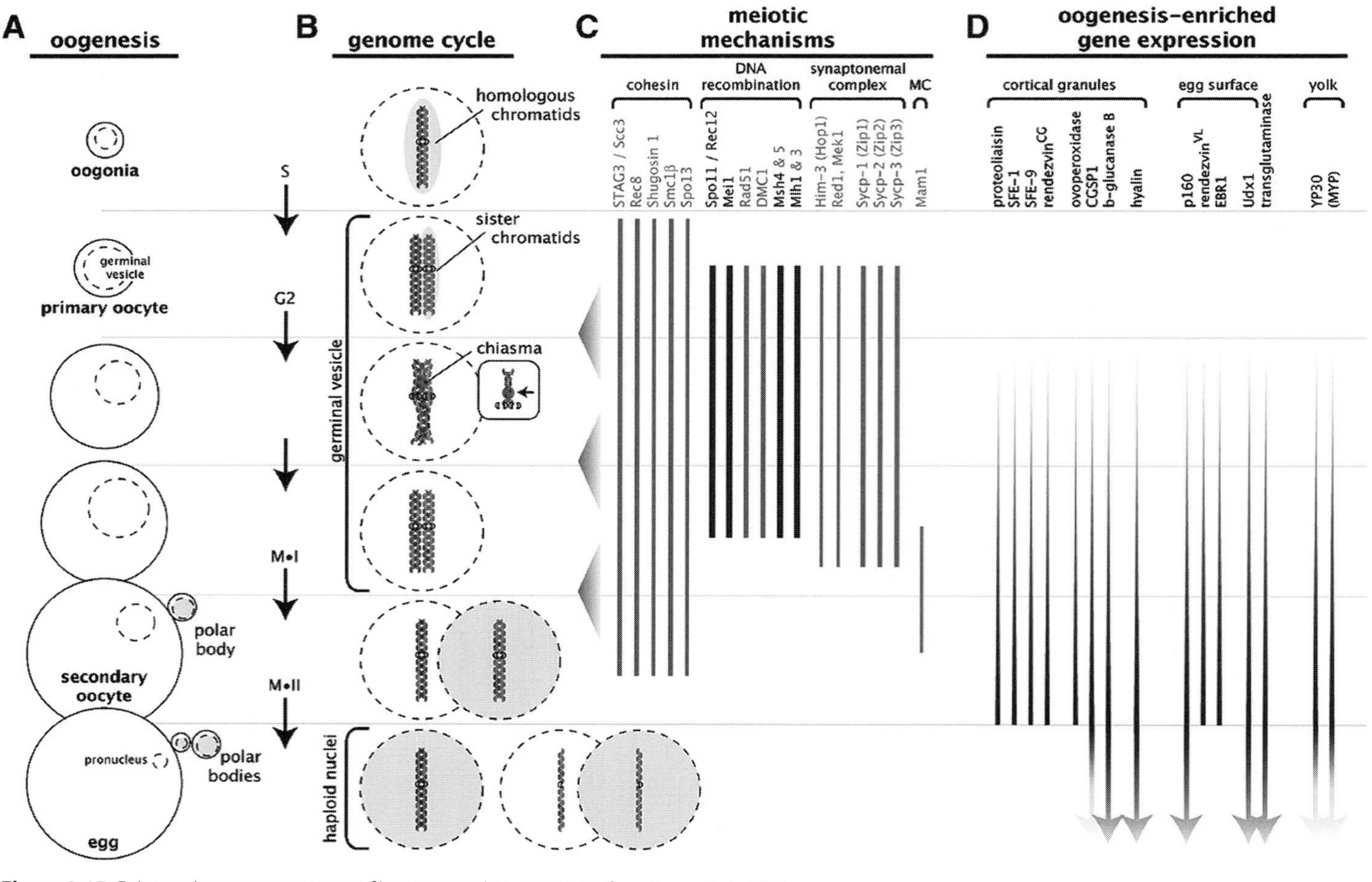

Figure 2.17 Enhanced gene expression profiles in sea urchin oogenesis (from Song et al. 2006).

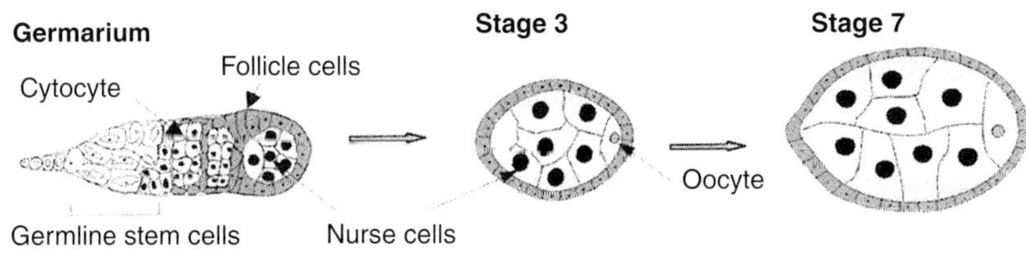

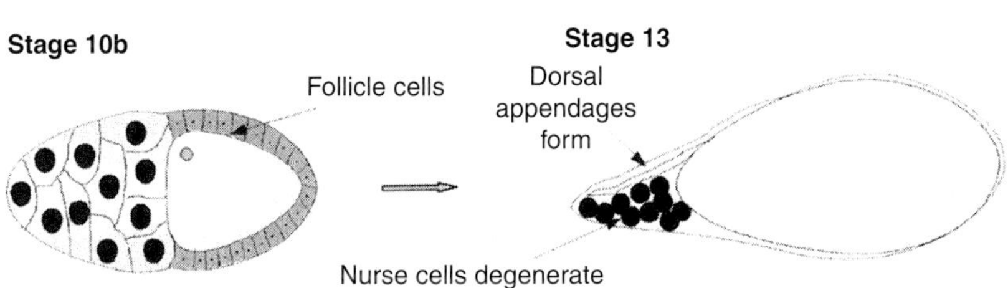

Figure 2.18 Main stages in *Drosophila* oogenesis showing the distribution of the nurse cells and follicles cells (from Trounson and Gosden 2003).

chromosome diminution while the germ cells maintain the full chromosome number. The genital primordia consist of two germ cells and two epithelial cells. After division of the peripheral germ cells, some of the cells destined to become spermatocytes undergo meiotic divisions and progress to the lumen. In other species, the primordial cells give rise by mitotic and amitotic divisions to packets or syncytial masses of 128 cells, each retaining cytoplasmic continuity with all the surrounding cells. The last stage of sperm maturation is often completed in the female tract and consists of the formation of pseudopods for motility and fusion of the cytoplasmic membranous organelles with the sperm plasma membrane and the consequent release of their lumen into the female cavity. In *Ascaris,* the large refringent body is formed at this time (Figure 2.19). This phase of maturation may be considered analogous to capacitation in the mammals. Nematode sperm remain vital in the female tract for up to months. External Na^+ and K^+ are both required for nematode sperm activation.

The female nematode *Ascaris suum* may produce 1.6 million oocytes per day and has a lifespan of one year. Little is known about oogenesis in this phylum, but in many species it is thought that the female germ cells remain as primary oocytes in the ovary and undergo maturation following fertilization. *Ascaris* oocytes accumulate large amounts of yolk, protein and carbohydrates during growth by intracellular synthesis and also incorporation by phagocytosis. There is no evidence that oocytes are connected by intercellular bridges during development. The oocyte synthesizes the primary oocyte envelope, which consists of three layers: an inner vitelline layer, a chitinous layer and a third ascaroside layer. A further coat, the mammillated external coat, is a product of the uterine secretions and is called a secondary envelope.

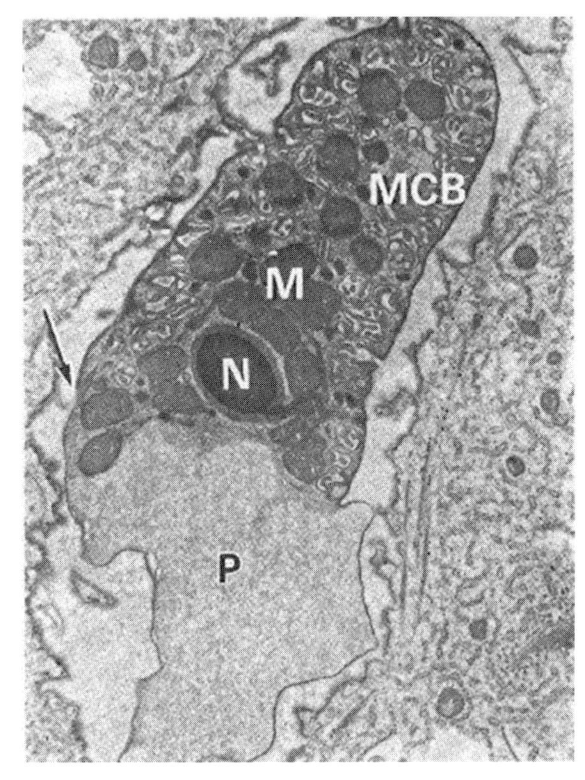

Figure 2.19 The refringent body in the activated spermatozoon of *Ascaris*. P – pseudopod (from Adiyodi and Adiyodi 1983).

Molluscs

Molluscs comprise the second largest phylum in the animal kingdom with 100,000 species and are either dioecious or hermaphrodite. The variety in sperm shape is enormous (Figure 2.20), however – different from other phyla – they are all uniflagellate. In the ovotestis of hermaphroditic species, we find female germ cells medially and male germ cells laterally. The Pelecypoda are dioecious, with the spermatogonia lying in the testis wall in small synchronous groups and in later stages are found in the lumen in 'sperm balls'. In each ball 250–2,000 spermatozoa are found with their heads directed towards the centre. In many bivalves, the testicular follicles are surrounded by nutritive somatic cells analogous to the nutritive phagocytes found in the sea urchin, while Sertoli cells are common in other species.

The common oyster, *Ostrea edulis,* is hermaphroditic. Individuals start life as males, then become females and finally males again. Although the sex changes are consecutive, both gametes may be found in the gonadal tube.

In the ovary of molluscs, the developing oocytes are closely associated with follicular cells, which increase in number by mitosis. In *Octopus vulgaris,* the oocytes are completely surrounded by follicle cells that form a syncytium. Follicle cells in molluscs serve many functions from nutrition, formation of the chorion and hormone production to the determination of the animal–vegetal and bilateral symmetries of the oocyte. The interaction of the follicular cells and the basal part of the oocyte may determine the localization of morphogenetic factors in this basal zone. In the prosobranch *Nassarius reticulatus,* the vegetal pole of the oocyte is morphologically distinct from the rest of the oocyte and is in fact the dorsal area of the cell that has been in contact with the follicle cells (Figure 2.21). The amount of yolk deposited in molluscan oocytes determines their relative size and the yolk may be both proteinaceous and lipid. The primary oocyte membrane, the vitelline membrane is deposited by the oocyte itself and is composed of proteins and polysaccharides similar to the glycocalyx of other animal oocytes. The chorion, formed in some species such as the squid, is also deposited while the oocyte is in the ovary, whereas the tertiary membrane found in decapods is formed by female accessory sex glands.

Cnidaria (Coelenterates)

The Cnidaria has three major classes, the Hydrozoa (hydroids), Scyphozoa (jellyfish) and Anthozoa (the

Figure 2.20 The range in sperm shape in the molluscs. The primitive type is shown in **(a)** (from Adiyodi and Adiyodi 1983).

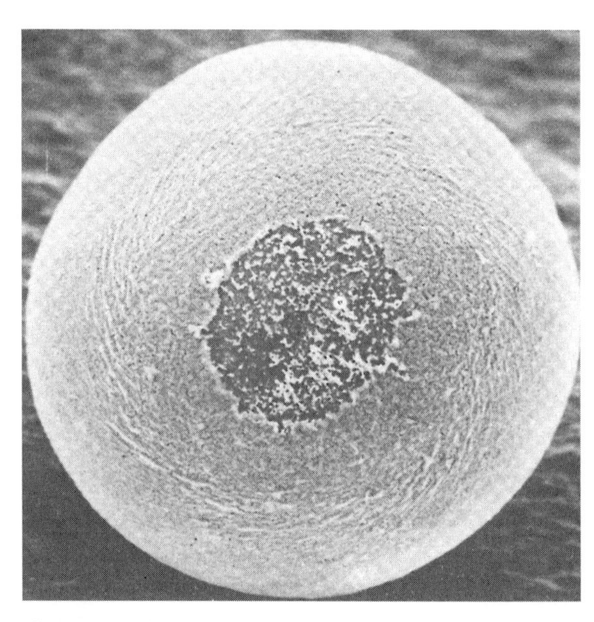

Figure 2.21 The polarized vegetal pole of oocyte of the mollusc *Nassarius* (from Adiyodi and Adiyodi 1983).

corals). Most of our knowledge has come from the former, the Hydroids. In the ovary, interstitial cells differentiate into oocytes. Some grow by cannibalizing others and become surrounded by a layer of follicle cells while meiotic reduction reduces the chromosome number from 12 to 6. In the hydrozoans, the germ cells originate from the epithelial cells of the ectoderm or endoderm depending on the species. Some of these epithelial cells 'dedifferentiate', becoming embryonic in nature producing spermatogonia of 8 μm. The cells then divide, forming a follicle, and the spermatogonia transform into the primary spermatocytes measuring 6.4 μm. The primary spermatocytes then divide by meiosis producing first two cells of 3.2 μm, the secondary spermatocytes, and then four spermatids of 2.3 μm. These meiotic cells are usually connected by cytoplasmic bridges. In spermiogenesis, the nucleus condenses and migrates to the blunt end of the cone. The middle piece and centrioles are now apparent, and the flagellum is about 20 μm long. There is no obvious acrosome, as is also the case for most teleosts. In hydrozoans, turnover of the germinal epithelium takes three days with a complete generation of mature sperm being released every 12 hours. The cycle from secondary spermatogonia to mature spermatozoa is about six days.

Chapter 3

The Players

There is much variation in the size and form of oocytes. Marine invertebrate oocytes are generally 60–150 μm in diameter, mammals about 100 μm, fish and amphibians about 1 mm, and we are all familiar with the sizes of bird eggs. Despite the differences in size and shape of oocytes across the animal kingdom, we may make some generalizations. First, oocytes are surrounded by several extracellular layers. The innermost structure, a fibrous glycoprotein sheet, is called the vitelline envelope in echinoderm and amphibians, zona pellucida in mammals and chorion in insects, fish, ascidians and molluscs. Externally, we find the extracellular matrix in mammals, composed of hyaluronic acid and cumulus cells; the jelly layer in echinoderms, annelids and amphibians; and the follicle cells in ascidians. All the extracellular coats, apart from the inorganic shells of reptile and bird eggs, which are deposited after fertilization, are present at fertilization. Therefore, the spermatozoon has to interact with and penetrate these layers before completing fertilization. The size of a spermatozoon is a mere fraction of that of the oocyte, often not exceeding one-millionth the volume of the oocyte; however, there is also an enormous variation in shape and size across the animal kingdom. Spermatozoa are often extremely long, reaching some 40 μm in sea urchins and mammals, 2–5 mm in some amphibians and 12 mm in insects. Although there is great variation in the shape and size of spermatozoa, we may regard them to be morphologically and functionally composed of four regions: the head, containing the nucleus and acrosome; the neck, containing the centrioles; the middle piece, containing the mitochondria; and the tailpiece or flagellum. In contrast to the oocyte, the spermatozoon has completed meiosis at the time of fertilization.

The Oocyte

The phylum Echinodermata is probably the most studied of all the phyla, with scientists for over 100 years adopting these marine creatures as an experimental model. Sea urchins provide vast numbers of gametes and have been the animal of choice with species such as *Paracentrotus lividus* (Figure 3.1), *Lytechinus variegatus* and *Strongylocentrotus purpurata* setting the scene. Even within a limited group such as the sea urchin, oocyte size varies between closely related species. For example, in the Californian echinoderm *Strongylocentrotus, S. purpurata* oocytes are 70–80 μm, *S. franciscanus* oocytes are 120–140 μm and those of *S. droebachiensis* are 140–160 μm. Thus in these animals, that occupy quite close ecological niches, oocyte volume can differ five-fold. In these species, oocyte size does not correlate with adult body size but seems to be a function of reproductive success. Sperm motility is inversely correlated with oocyte size, with those bearing large oocytes having slower swimming spermatozoa. It is thought that larger oocytes have a higher probability of being fertilized. Starfish oocytes are also extremely useful for fertilization studies because of their larger size and relative transparency (Figure 3.2).

The annelids are another interesting phylum, with over 14,000 species occupying most marine, freshwater and subterranean habitats and showing an enormous variation in gamete morphology and size. Annelid

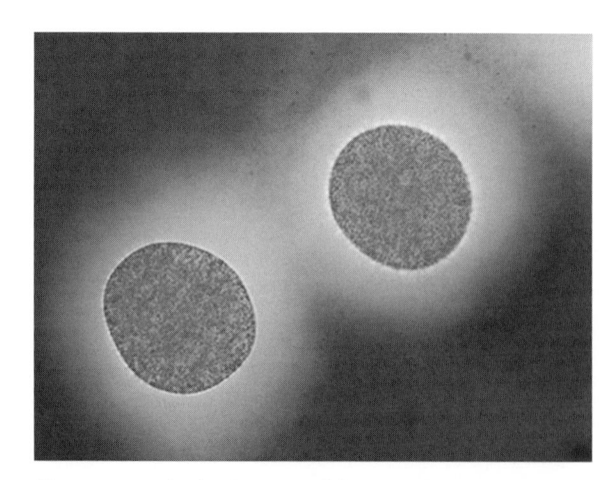

Figure 3.1 Unfertilized oocytes of the sea urchin *Hemicentrotus pulcherrimus* showing the extensive layer of jelly. The oocytes measure 80 μm in diameter (courtesy of Dr Keiichiro Kyozuka).

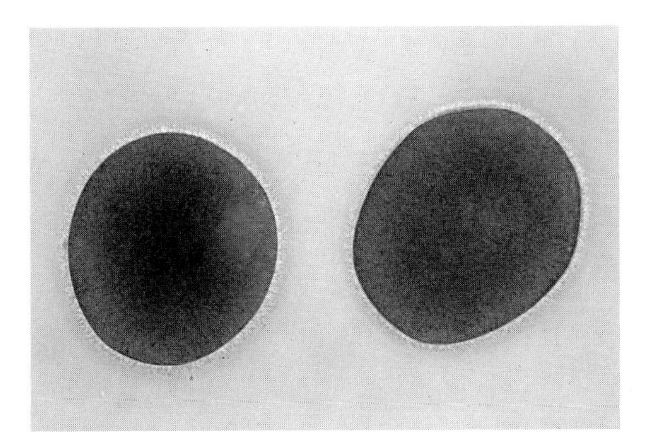

Figure 3.2 Oocytes of the starfish *Asterina pectinifera* 50 minutes after application of the maturing hormone 1-methyladenine (bottom). The oocytes are 160 μm in diameter. The jelly layer is much thinner and more compact than that of the sea urchins (courtesy of Dr Keiichiro Kyozuka). (A black-and-white version of this figure will appear in some formats. For the colour version, please refer to the plate section.)

oocytes range from 20–40 μm in *Polyophthalmus pictus* to 1.17 mm in *Paronuphis antartica* (Figure 3.3). Other commonly used oocytes are those of the ctenophore *Beroe ovata*, or comb jelly, which has a diameter of 1 mm and extremely clear cytoplasm. Ascidians, or tunicates, are in fact part of the sub-phylum Urochordata, since the larval stage demonstrates chordate characteristics. These animals are hermaphroditic and the gametes are self-sterile. Oocytes vary in size from 110 μm in *Molgula* to 730 μm in *Polycarpa tinctor*. The ascidian oocyte lends itself to studies of fertilization since the chorion, consisting of a fine meshwork of fibres, may be manually removed to generate a nude oocyte. Outside the chorion, most species are surrounded by a layer of follicle cells which allow the oocyte to float (Figure 3.4).

The class Insecta, with over 750,000 species, is without doubt the most successful animal group and in fact is larger than all the other groups put together. Insect

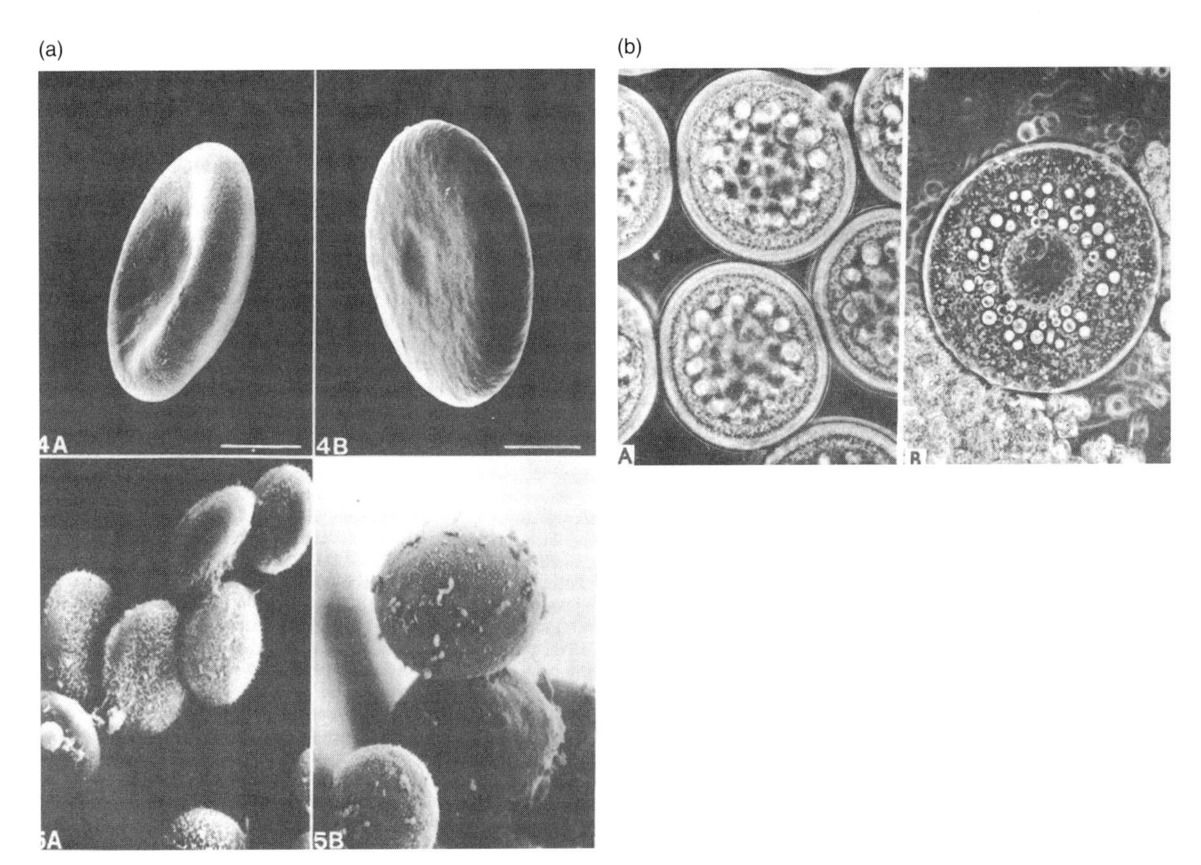

Figure 3.3 **(a)** The two top frames show oocytes of the marine worm *Arenicola marina* that change shape from a concave form at prophase1 to a disc shaped form at metaphase 1 following spawning. The bottom two frames show Discoid prophase oocytes of *Pectinaria gouldii* and spherical mature oocytes of the same species. **(b)** The left frame shows mature oocytes of the free spawning nereid *Nereis grubei* with abundant lipid and jelly precursor material, and the right frame shows oocytes of *Nereis limnicola*, which has less jelly precursor material and lipid (from Adiyodi and Adiyodi 1983).

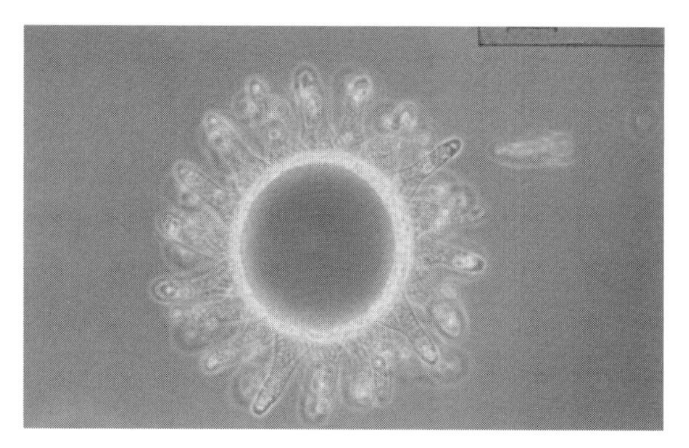

Figure 3.4 An oocyte of the ascidian *Ciona intestinalis* showing the follicle cells, which help in flotation, and the thick extracellular coat called the chorion. (A black-and-white version of this figure will appear in some formats. For the colour version, please refer to the plate section.)

Figure 3.6 Grooves in the chorion of fish oocytes guide the spermatozoa to the micropyle. *Lucio cephalus* **(a–c)** and *Sturisoma* **(d and e)**. Scale bars A-500 μm, B-10 μm, C-100 μm, D-500 μm and E -50 μm (from Tarin and Cano 2000).

oocytes are extremely large, reaching several millimetres, and are surrounded by a thick multilayered chorion deposited by the ovarian follicle cells. A perforation in the chorion, the micropyle, allows sperm penetration. The micropyle of several insects studied to date has been shown to be covered by a gelatinous cap (Figure 3.5). Fish oocytes may also reach 2 mm in diameter and are enclosed in a thick chorion with a single micropyle for sperm penetration located at the animal pole. The chorion is also in turn embedded in a jelly layer and extracellular matrix (Figure 3.6).

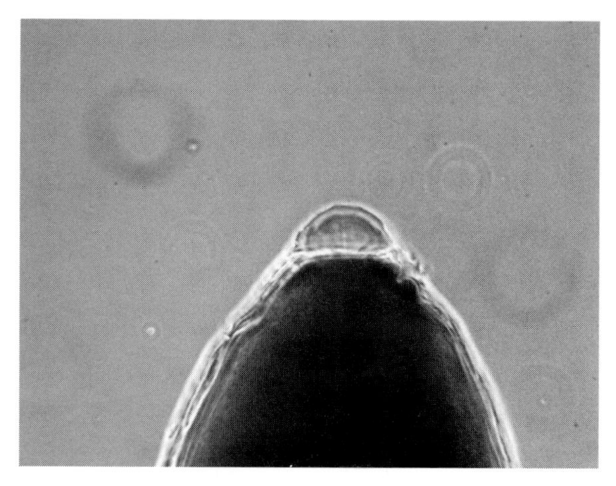

Figure 3.5 The micropyle and jelly cap in the oocyte of the housefly *Mus domestica* (courtesy of Professor Ryuzo Yanagamachi). (A black-and-white version of this figure will appear in some formats. For the colour version, please refer to the plate section.)

Amphibian oocytes range from 1.8 mm diameter in *Discoglossus pictus* to 4.4 mm in *Flectonotus pygmaeus*. They are ovulated from the ovaries into the body cavity with a thin vitelline envelope of several μms secreted by the oocyte itself. The molecular structure and composition of this envelope is modified as the oocytes pass through the pars recta. As the oocytes pass through the posterior portion of the oviduct, secondary envelopes or jelly coats are laid down around the vitelline envelope, completing and conferring fertilizability to the oocyte (Figure 3.7).

Mammalian oocytes range in size from 100 μm in eutherian mammals to the large, yolky oocytes of the monotremes, which reach 4 mm. Both marsupial and monotreme oocytes differ from eutherian oocytes in that they are surrounded by a shell. There are only three species of monotreme, with the Australian platypus and spiny anteater being the most recognized, while there are over 300 marsupials. The vast majority of mammals are eutherian mammals, which amounts to 6,000 species. The typical eutherian oocyte is approximately 100 μm in diameter, is covered by a glycoprotein sheet several μms thick and is embedded

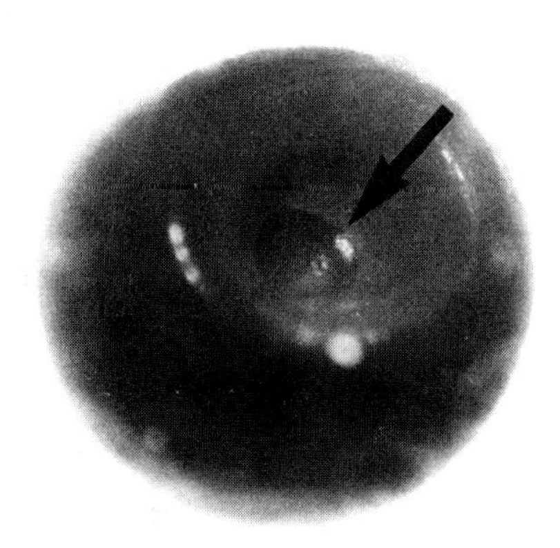

Figure 3.7 An oocyte of the painted frog *Discoglossus pictus* showing the dimple, the site of fertilization (courtesy of Dr Chiara Campanella).

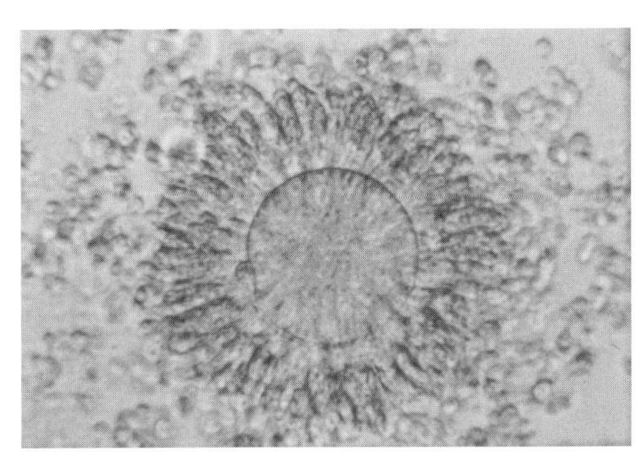

Figure 3.8 The metaphase 2 human oocyte, showing the polar body to the left, the zona pellucida, and the extensive layer of cumulus oophorus, which is made up of cells and hyaluronic acid. (A black-and-white version of this figure will appear in some formats. For the colour version, please refer to the plate section.)

in an extracellular matrix of cumulus cells and hyaluronic acid. Over the past 30 years, human in vitro technology has given scientists the opportunity to study human oocytes (Figure 3.8).

For many decades it has been the general consensus that the fertilizing spermatozoon penetrates the oocyte at any location. In amphibians for example, the spermatozoon preferentially penetrates the animal hemisphere where the metaphase 2 chromosomes are located. In some animals, such as fish, insects and cephalopods, the extracellular oocyte coat has a narrow canal – a micropyle – through which spermatozoa enters the oocyte. However, even here the micropyle is filled with a gelatinous mass that needs to be navigated by the spermatozoon. We will see in the following chapters that perhaps there are preferential entry sites in the oocytes of all animals.

We saw in the previous chapter that oocytes at ovulation are arrested at various stages of meiosis, ranging from prophase 1 in molluscs, metaphase 1 in ascidians, and metaphase 2 in mammals and amphibians to the completion of meiosis in sea urchins (see Figure 2.6).

The Spermatozoon

The spermatozoon of the painted frog *Discoglossus pictus* is 2500 μm long with a nucleus that is 700 μm long, while in the fruit fly *Drosophila bifurca,* the spermatozoon, coiled in a ball, would measure more than two inches if straightened out. It has been suggested that this extreme length is an adaptation to the long

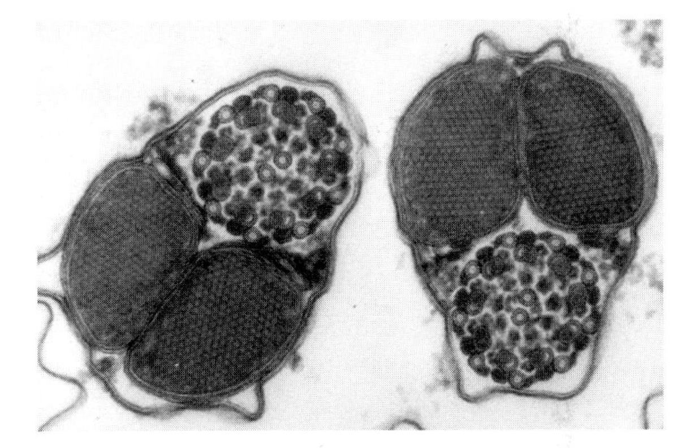

Figure 3.9 Cross-sections of the flagella of the spermatozoon of the medfly (*Ceratitis capitata*) showing the 9+9+2 axoneme design common to many insects. Two mitochondrial derivatives showing crystallization can be seen (courtesy of Dr Romano Dallai).

reproductive tract of the female, and it is thought that longer sperm cells are better than shorter sperm at displacing competitors from the female's seminal receptacle. The insects, being the largest group in the animal kingdom, have no counterpart in their capacity to adapt to different environments. There is an enormous variation in the design of insect spermatozoa, with structures ranging from conventional flagellated sperm to aflagellated immotile cells. Of particular interest is the insect sperm axoneme, which in many cases has an extra set of microtubules outside the standard 9+2 format, described as the 9+9+2 axoneme (Figure 3.9). These accessory tubules may make the axoneme more resistant to damage, or they may be part of the motor system. Others have suggested that the extra tubules are a storage site for polysaccharides.

The annelids also demonstrate a wide variety of sperm shape from the simple round primitive shapes to filoform shapes (Figures 3.10 and 3.11) which seems to be correlated with the mode of reproduction. Franzén in 1956 suggested that primitive sperm with a simple acrosome, spherical nucleus, a small number of mitochondria and a free flagellum were typical of

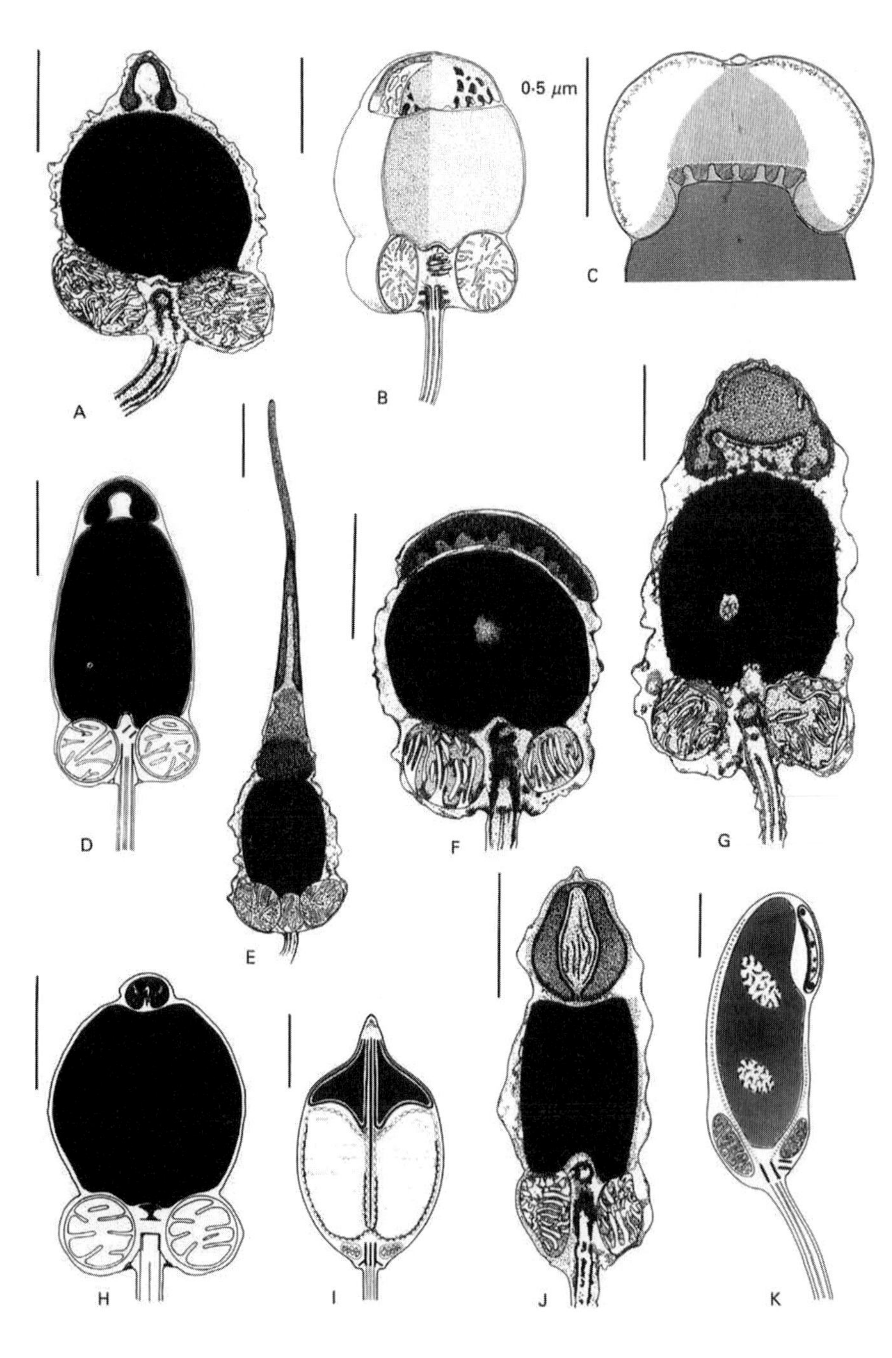

Figure 3.10 Longitudinal sections of polychaete ect-aquasperm (previously called 'primitive sperm') and in (K) an example of ent-aquasperm (from Jamieson and Rouse 1989).

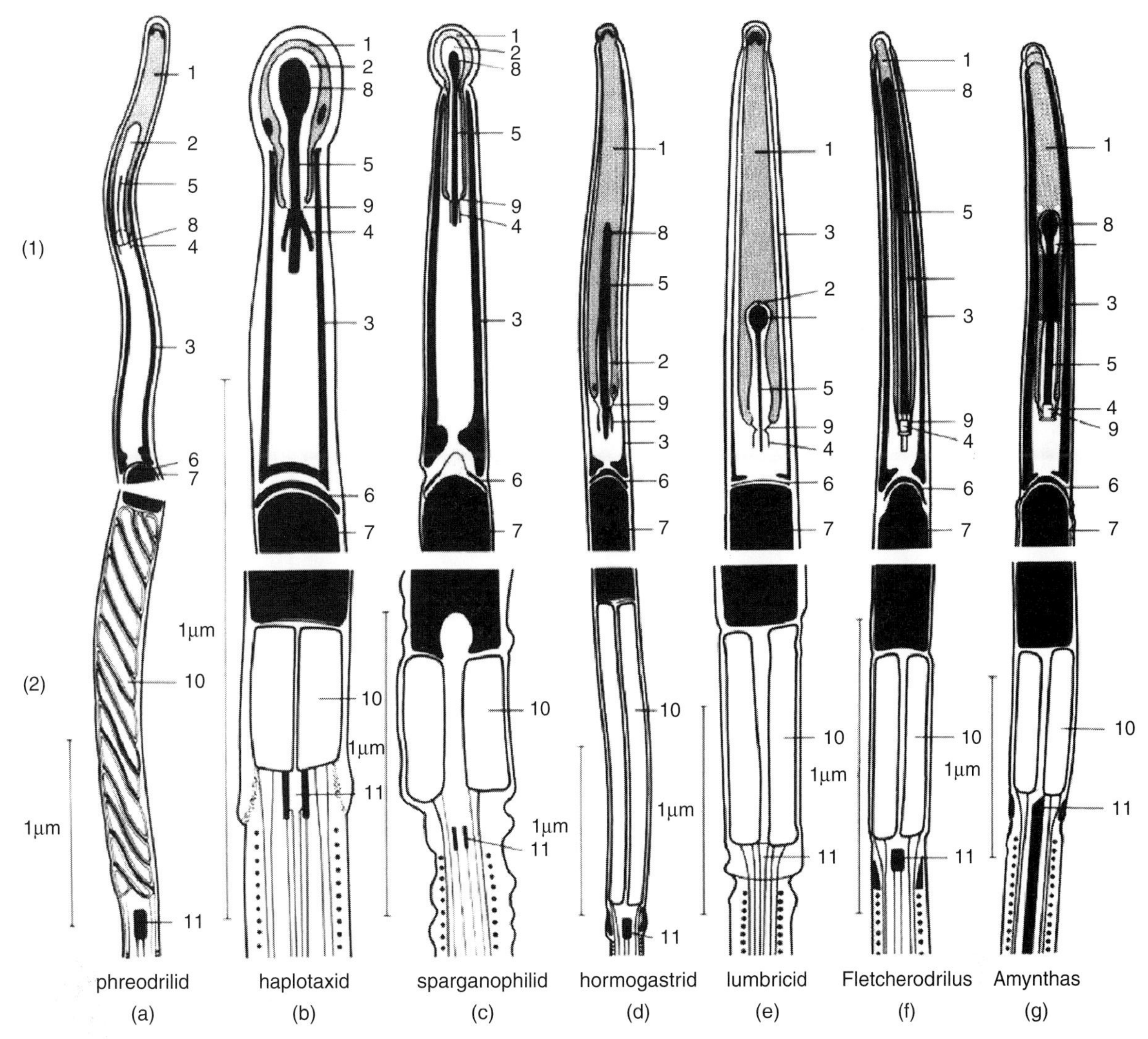

Figure 3.11 (1) Different forms of oligochaete spermatozoa and (2) Transmission electron micrographs (TEMs) of the oligochaete sperm *Eudrilus* (courtesy of Barrie Jamieson).

animals practising external fertilization, while Rouse and Jamieson in 1987 classified annelid sperm based on their function:

Ect-aquasperm: the sperm are released into the water, as are the eggs, and the fertilization takes place in the water.

Ent-aquasperm: the sperm are released into the water but then are gathered into the female and often stored in spermathecae before fertilizing the eggs.

Introsperm: the sperm have no contact with the water and pass directly from the male to the female.

The Mollusca is the second largest invertebrate phylum, with over 100,000 species described. It covers all marine, freshwater and terrestrial

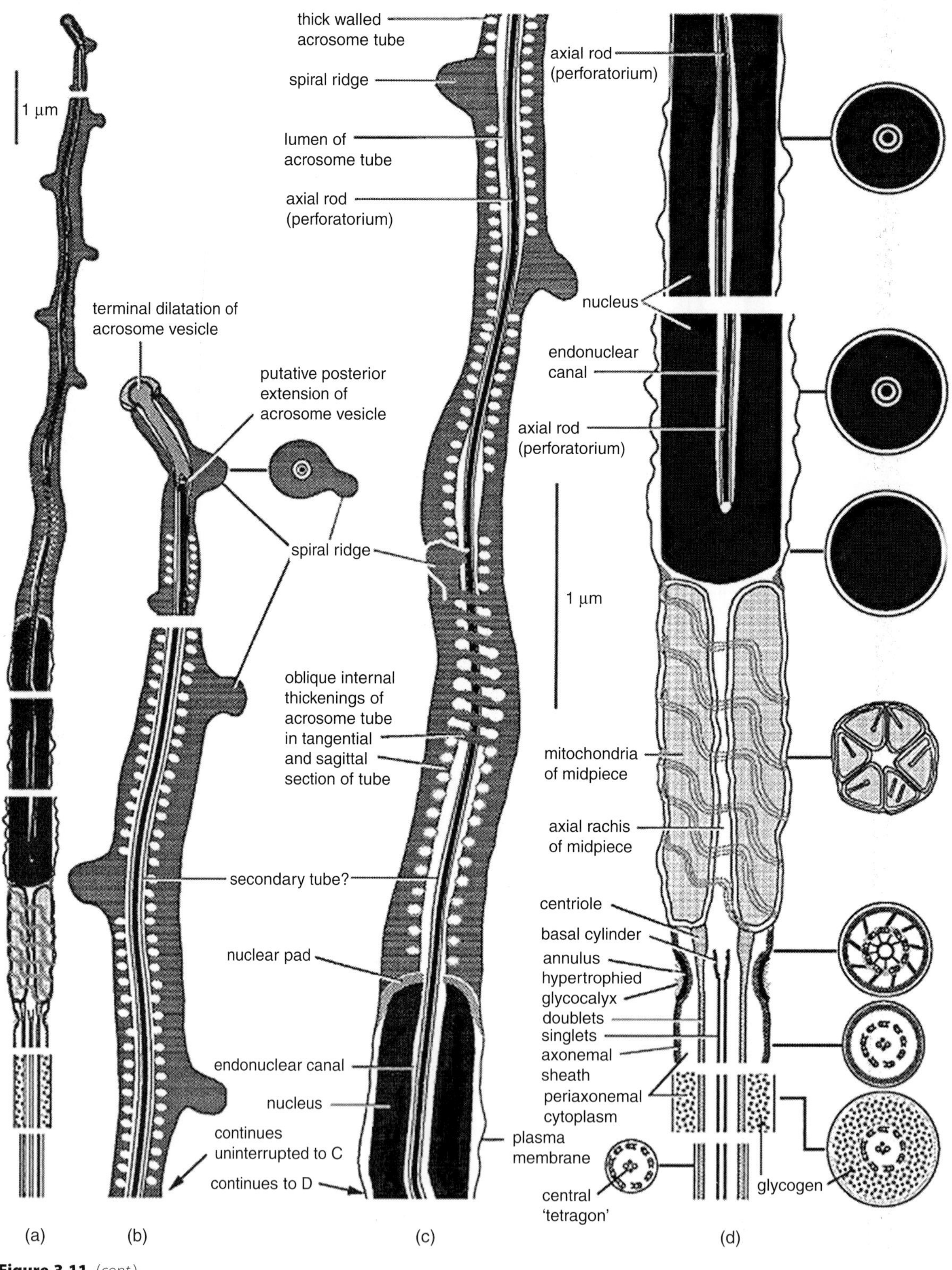

Figure 3.11 *(cont.)*

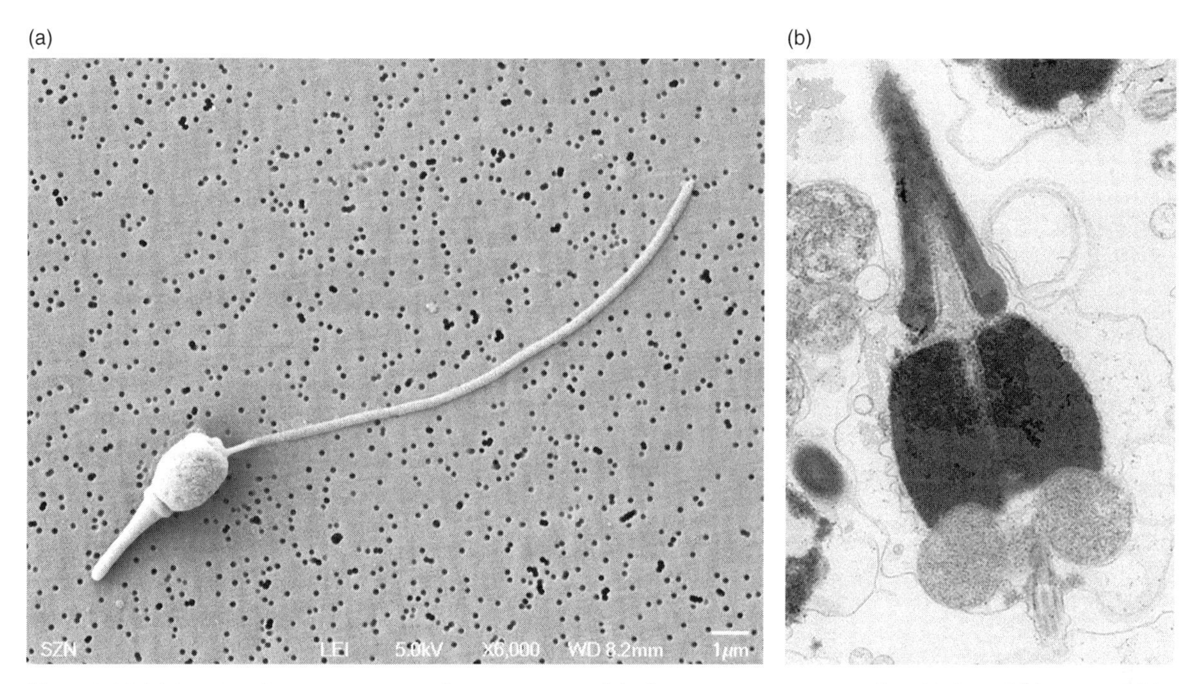

Figure 3.12 **(a)** Scanning electron microscope of a spermatozoon of the free spawning common mollusc *Mytilus* and **(b)** a section of the same spermatozoon under the transmission electron microscope. Note the primitive structure of the apically located small acrosome, the conical nucleus and the round mitochondria in the mid-piece (courtesy of Dr Elisabetta Tosti).

environments and spans many reproductive strategies from free spawning to internal fertilization following copulation. Bivalves such as oysters and mussels are free spawning, and the cephalopods undergo internal fertilization after the transfer of sperm by copulation or pseudocopulation with spermatophores. Sperm design reflects reproductive strategy, with the free spawning bivalves having primitive sperm (Figure 3.12). Those with internal fertilization have modified sperm. For example, octopus sperm has a very particular design consistent with copulation (Figure 3.13). Molluscs show the greatest diversity in sperm morphology throughout the animal kingdom (see Figure 2.20), with the modified type occurring mainly in the gastropods and cephalopods. Primitive molluscan spermatozoa have an anteriorly located acrosome of 1–2 μm in diameter. At the base of the nucleus are usually four mitochondria with two centrioles found in the midline between. The mollusc flagellum composed of the standard 9+2 microtubular structure reaches some 40–50 μm. In octopus spermatozoa the acrosome surmounts the nucleus and is helical in structure. There is only a single centriole in octopus sperm. Pulmonata, which include the common garden snail *Helix aspersa* and the common freshwater snail *Lymnaea stagnalis*, also have modified spermatozoa characterized by a small nucleus, a small acrosome and a long tail that may reach 1700 μm in some species. All molluscan spermatozoa, whether of the primitive type or modified type, have abundant polysaccharide deposits found at various sites from around the microtubules of the flagellum to between the mitochondria in the middle piece.

Cnidarian spermatozoa are similar to teleost sperm in that they lack an effective acrosome, although there are periodic acid-Schiff (PAS)-positive vesicles that may serve the same purpose which are usually found at the apical end of the nucleus. The head of cnidarian sperm is typically

3 μm long and about 1 μm in width with a middle piece measuring 1x1 μm and a tail of 30–40 μm long. The middle piece consists of a few mitochondria arranged around the centrioles.

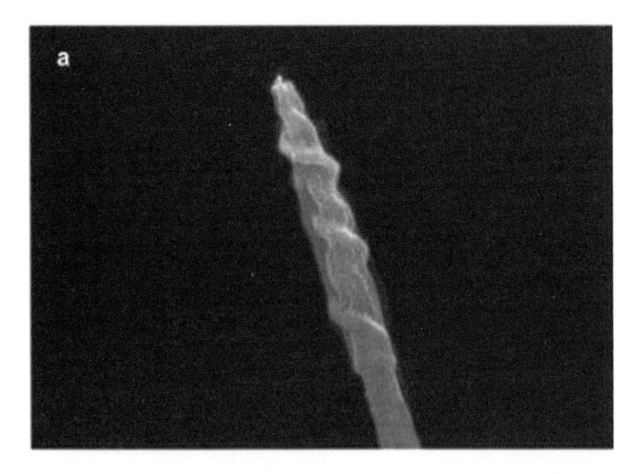

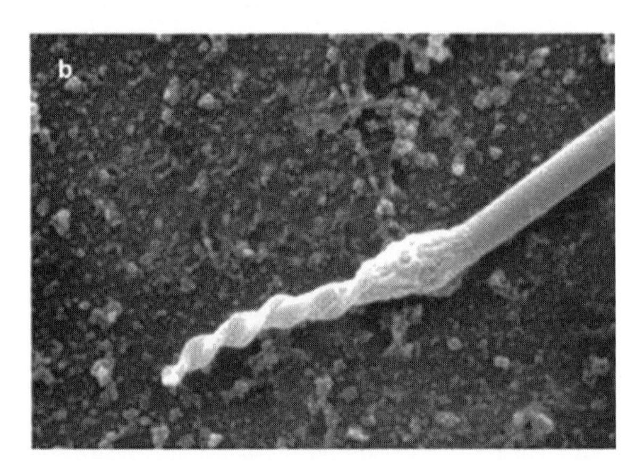

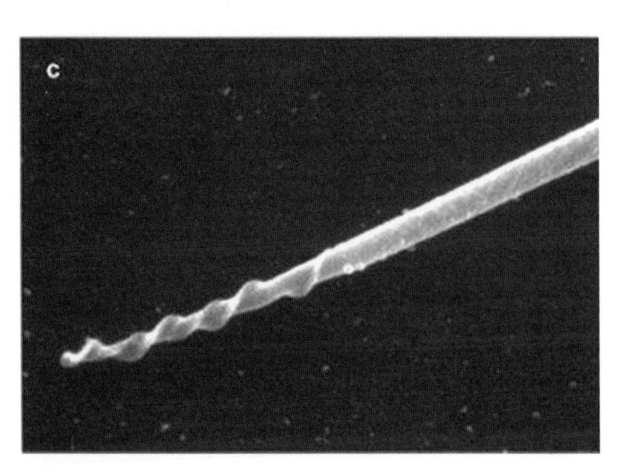

Figure 3.13 In contrast to the bivalves, molluscs that practice internal fertilization, such as *Octopus vulgaris*, have a complex helical sperm structure (courtesy of Dr Elisabetta Tosti).

Nematode spermatozoa do not possess flagella or acrosomes and move by amoeboid movement. The anterior portion is clear, and the posterior portion contains the major organelles, which include the nucleus, the mitochondria and lipid-like refringent bodies typical of *Ascaris*. The plasma membrane of these amoeboid cells is particular in that it is not uniform over the entire surface of the spermatozoon. The anterior pseudopod membrane does not have pores and demonstrates plasticity since it is this part of the cell that extends and retracts, while the plasma membrane surrounding the main body is associated with a dense glycocalyx and often contains pore-like areas (see Figure 2.19). There are more than 100 mitochondria in spermatozoa of *Ascaris*, whereas in *Dioctophyma* there are none. Since *Ascaris* is found in anaerobic environments and the mitochondria lack cristae, it appears the mitochondria serve purposes other than energy production, for example in the storage of intracellular calcium. Although nematode spermatozoa do not contain flagella, centrioles are present in many species and have an unusual structure not displaying the typical nine triplet structure. In the Ascarids, the

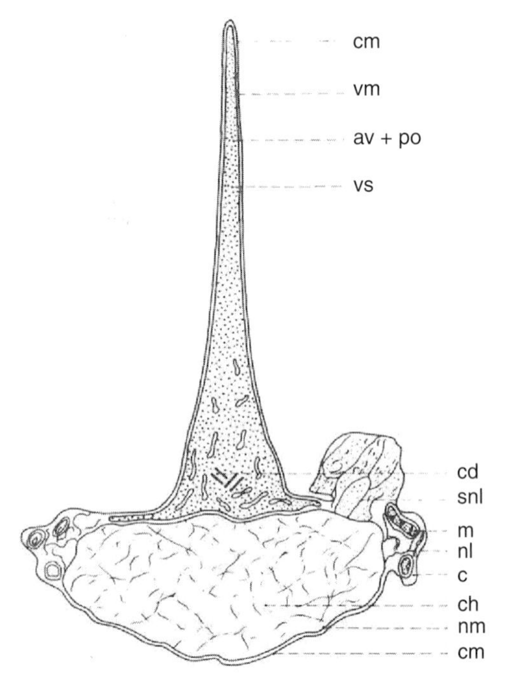

Figure 3.14 A drawing to show the flagella-less spermatozoon of a crustacean decapod, *Natantia* (from Pochon-Masson 1968).

outstanding feature of the sperm cell is the refringent body in the posterior cytoplasm made up of lipids and proteins.

Crustacean spermatozoa are either flagellate or have strange divergent shapes and are immotile. Ostracods have one of the longest spermatozoa in the animal kingdom. The spermatazoa can be several millimetres in size, often longer than the body of the male; however, it lacks a flagellum. Decapod spermatozoa, which include crabs, shrimps and lobsters, are spherical, also devoid of a flagellum and have an acrosome formed from ergastoplasmic cisterns which act as a substitute for the Golgi (Figure 3.14).

Most fish spermatozoa lack an acrosome (Figure 3.15); exceptions are the sturgeon and the paddlefish, whose oocytes are also different due to having several micropyles. Generally speaking, spermatozoa from fish have a simple or primitive structure, with a small roundish head (2–4 μm), from 2 to 9 mitochondria in the middle piece and a typical 9+2 arrangement in the flagellum. The flagellum is however quite unique in having fins, giving the appearance of a ribbon, and varies in length from 20 to 100 μm. The spermatozoa in sturgeons and paddlefish have an elongated head to accommodate the acrosome measuring 2–10 μm.

There is much information on the morphology of avian spermatozoa (Figure 3.16), which is often described as sauropsid, since the more primitive non-passerine spermatozoon is similar to that of the reptiles. In the non-passerine domestic fowl, the acrosome is 2 μm long with the nucleus being 12 μm long and the flagella 80 μm long. Avian spermatozoa have been instrumental in studying sperm motility. In 1888, Ballowitz demonstrated by light microscopy that the rooster sperm flagellum could be dissected into 11 longitudinal fibres. In the 1950s, with the advent of electron microscopy, it was shown that the flagella was composed of two central singlet microtubules surrounded by 9 doublet microtubules. The 9+2 axoneme structure was discovered. The evolution of vertebrate spermatozoa is a subject in its own right, and the reader should consult classical texts for further details (for example by Jamieson 1999).

In the mammals, the sperm nucleus is flattened. Although the process of spermatogenesis is essentially similar in marsupials and eutherians, in the former,

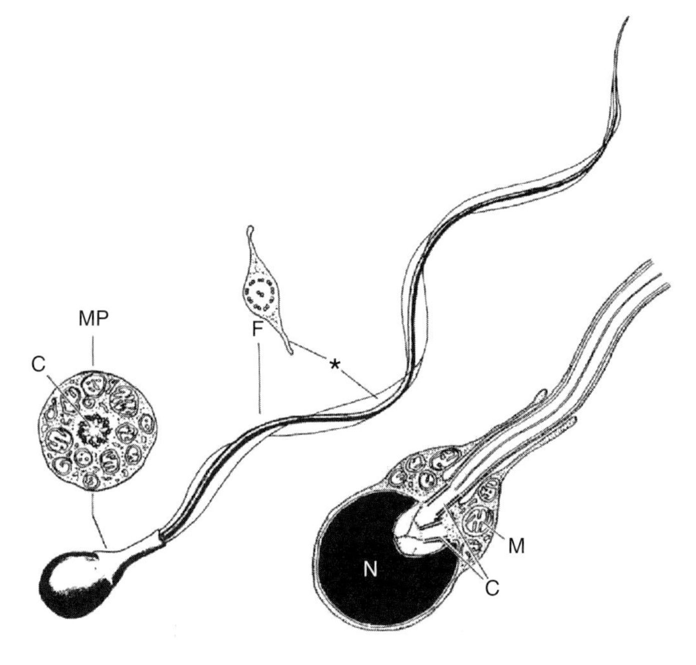

Figure 3.15 Most teleost fish sperm lack an acrosome, which may be correlated with the fact that the spermatozoa enter the oocyte through a preformed canal the micropyle. This diagram represents a spermatozoon of Oryzias latipes. C – centriole, N – nucleus, M – mitochondria. Note the ribbon-like sperm flagellum (from Tarin and Cano 2000).

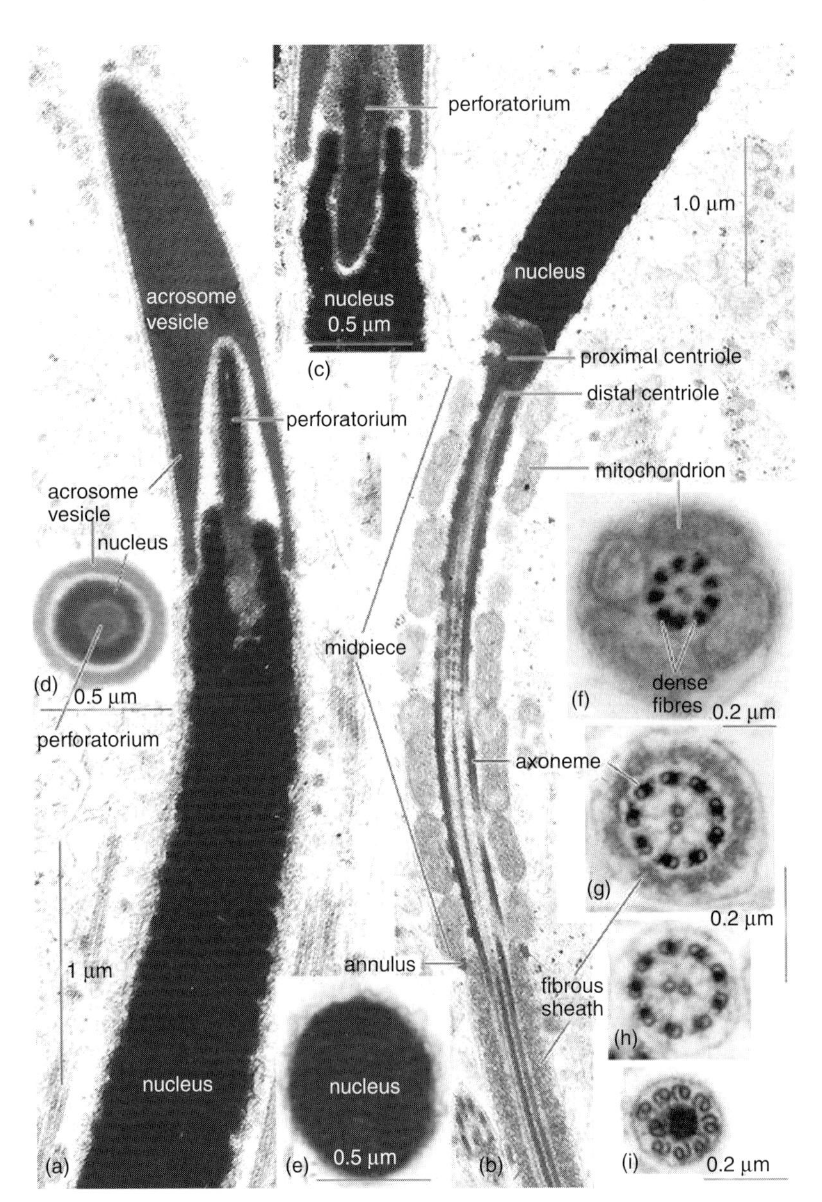

Figure 3.16 TEMs of the spermatozoon of the rooster *Gallus gallus* showing the perforatorium and the acrosome (courtesy of Barrie Jamieson).

flattening occurs at 90° to the tail while flattening in the eutherians is parallel to the tail (Figure 3.17). Therefore, the acrosome forms a cap in the eutherians, but it remains a disc-like structure in the marsupials. In contrast, monotreme sperm are non-mammalian in character, resembling the sperm of birds and reptiles. Eutherian sperm vary considerably in the size and topographical location of the acrosomal cap (Figure 3.18). The human spermatozoon is probably the most studied of all species to date, and detailed holographic microscopy shows that the mean head length is 5 μm, the width is 3.5 μm and volume is 14 μm. Modern microscope techniques such as 3D holography and Raman spectroscopy have revealed much information on the structure and function of human spermatozoa.

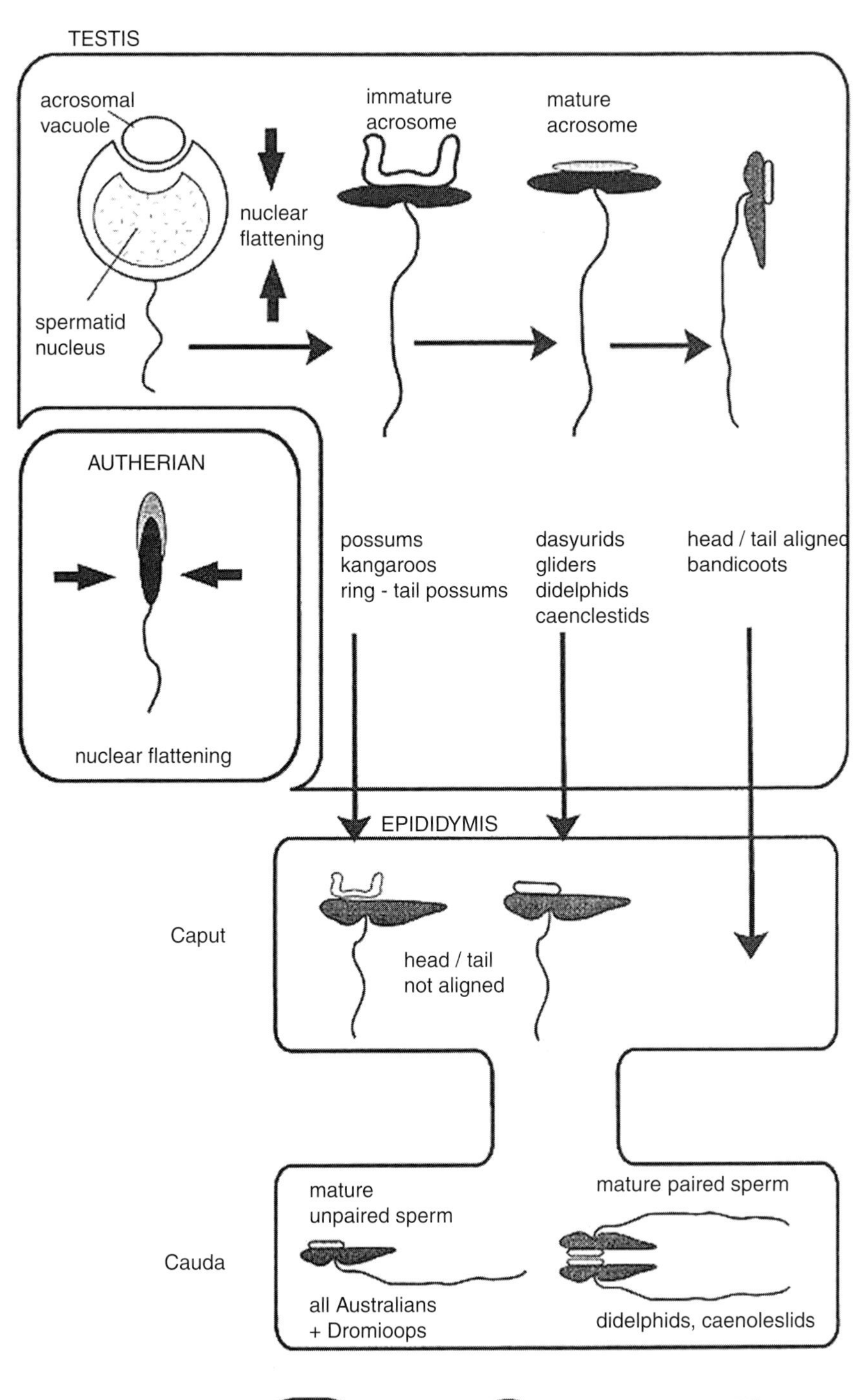

Figure 3.17 Spermiogenesis and epididymal maturation of spermatozoa in marsupials showing the position of the acrosome (from Tarin and Cano 2000).

Acrosome cap
Equatorial segment of acrosome
Postacrosomal region
Bull
Human
Mouse

Figure 3.18 The size, shape and location of the acrosome in spermatozoa of the bull, the human and the mouse. The sperm heads are approximately 9 µm, 4.5 µm and 7 µm, respectively.

Chapter 4

Sperm–Oocyte Interaction

Both spermatozoa and oocytes are maintained in the gonads in a quiescent state. Following their release into their reproductive environments, they adapt to a new set of conditions. These pre-activation events prepare the gametes for their imminent interaction together. The spermatozoon will then encounter a series of signals emitted from the female gamete, starting with chemotactic signals, followed by its interaction with the outer extracellular matrix and then the vitelline envelope. At each step, the spermatozoon undergoes a change in physiology which is a prerequisite that prepares it for the successive interaction. The crucial event is the acrosome reaction. Spermatozoa that do not respond to a correct sequence of signals are halted in their progression by defective signalling and fall to the wayside. It should be pointed out that there is considerable overlap in these events, and it is not often clear when an activation process starts and when it ends, whether it be the initiation of motility, chemotaxis or the acrosome reaction. In mammals at least, we may consider several steps; loose attachment of the spermatozoon, firm adhesion, induction of acrosomal exocytosis and penetration of the zona pellucida. Oocytes also undergo pre-activation events when ovulated, and if the oocyte is not activated by a competent spermatozoon at the correct moment in time, these physiological changes will increase in magnitude, leading to oocyte aging. In the last stage of gamete interaction, i.e. the fusion of the plasma membranes of the two competent gametes, the oocyte is triggered to undergo a rapid, polarized and highly synchronized reorganization at the structural, physiological and molecular levels. This is oocyte activation, which sets the scene for the first stages of development.

Release of Gametes into the Environment and the Pre-Activation of Spermatozoa

In aquatic animals, the spermatozoa are maintained in a repressed metabolic state in the testis by the concentration of ions, the pH or osmolality of the gonadal environment. For example, sea urchin spermatozoa are kept immotile in the testis owing to the high CO_2 tension in the semen that keeps the intracellular pH at about 7.2. When exposed to the different ionic milieu of the seawater at spawning, sperm motility and oxygen consumption is activated by exposure to the higher pH of the seawater, which leads to the activation of a Na^+/H^+ exchanger and release of acid (H^+) by the spermatozoon into the environment. The resulting increase in intracellular pH of the spermatozoon cytoplasm to pH 7.4 triggers the hydrolysis of adenosine tri-phosphate (ATP) by the mitochondrial oxidation of fatty acids and this is thought to activate dynein, the motor component of the axoneme responsible for sperm flagellar movement. Dynein, an ATPase, is inactive below a pH of 7.3. After spawning, the mitochondrial activity of sea urchin spermatozoa increases 50-fold. The intracellular pH increase in sea urchin spermatozoa at spawning is Na^+ dependent and is modulated by Zn^{2+}. It also depends on extracellular K^+, which is lower in the seawater than the raw semen. Therefore, spawning could hyperpolarize the membrane potential, which lies between -36 and -56 mV, thus triggering the voltage dependent Na^+/H^+ exchanger (Figure 4.1).

Trout spermatozoa are maintained quiescent in the testis by the high external K^+ of the semen. At spawning, the lower K^+ concentration of the external environment causes an immediate but transient increase in cAMP that triggers the phosphorylation of axonemal proteins leading to sperm motility

(Figure 4.2). The decrease in external K^+ also leads to a hyperpolarization which is independent of the intracellular pH of the spermatozoon. K^+ channel blockers such as TEA, and Ca^{2+} channel blockers, such as verapamil, block sperm motility in the trout.

In marine teleosts, sperm motility is triggered by exposure to the hypertonic milieu of the seawater and seems to involve a change in intracellular pH and intracellular calcium, whereas in some freshwater fish, sperm motility is triggered by the hypotonic milieu of the external environment. Even in brackish-water fish, gametes are seldom laid in isotonic water. In general, the fertile lifespan of marine and freshwater fish spermatozoa is rather short, lasting less than a few minutes. In the frog, *Discoglossus pictus* sperm motility is triggered at spawning following exposure to hypo-osmotic water.

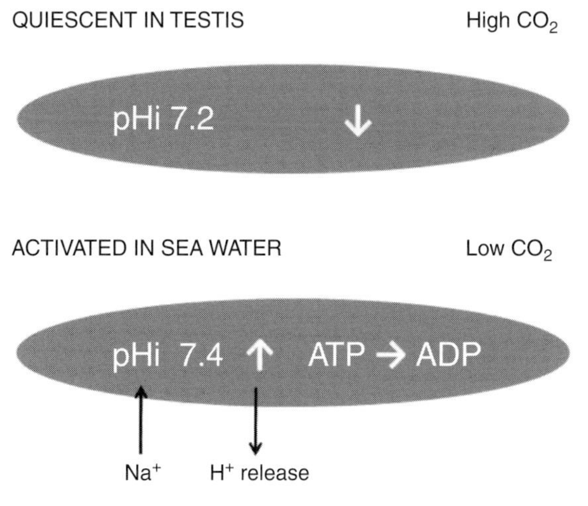

Figure 4.1 Ionic pathways involved in activation of sea urchin spermatozoa as they are released into the environment.

Mammals

Mammalian spermatozoa attain motility while being transported through the epididymis in a process called *maturation*. Ion concentrations vary considerably in the epididymis with Na^+ decreasing from 100 mM in the caput to less than 50 mM in the cauda, while K^+ increases inversely from 20 to 40 mM. It is thought that these ion gradients trigger, in part, the maturation process. Since the mouse sperm membrane potential is driven by K^+ and increasing external K^+ can depolarize the membrane, this may lead to the opening of voltage-dependent calcium channels and changes in intracellular Ca^{2+}. One of the major factors involved in sperm motility is an increase in intracellular cAMP acting through a protein kinase A that regulates the regulatory protein associated with the axoneme. Although all the necessary morphological structures

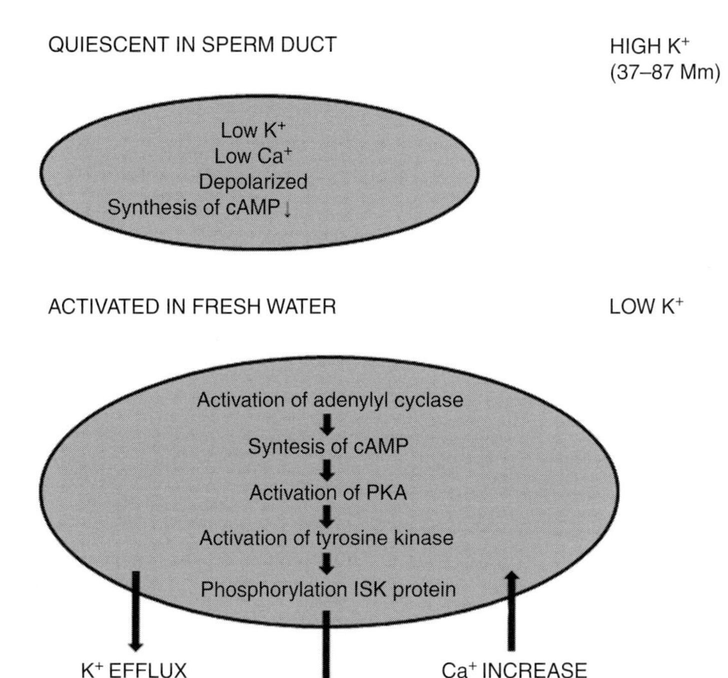

Figure 4.2 The mechanism of activation of trout spermatozoa.

for flagellar activity are assembled during spermiogenesis, testicular spermatozoa are essentially motionless, even when washed and placed in a physiological solution. Sperm from the caput epididymis begin to display motility, and by the time they reach the cauda, they are capable of full progressive forward motility. The ability to move is probably regulated at the level of the plasma membrane, since de-membranation and exposure to ATP, cAMP and Mg^{2+} triggers movement. Transfer of a forward motility protein and carnitin from the epididymal fluid are believed to be important for the development of sperm motility. Since the osmolality and chemical composition of the epididymal fluid varies from one segment to the next, it may be that the sperm plasma membrane is altered stepwise as it progresses down the duct, and motility is controlled by the interplay between cAMP, cytosolic Ca^{2+} and pH_i. During maturation, the spermatozoa use up endogenous reserves of metabolic substrates, becoming dependent on exogenous sources such as fructose, at which point they shed their cytoplasmic droplet.

Lipids and proteins in the plasma membrane of spermatozoa undergo distinct changes during epididymal maturation, with cholesterol being one of the key molecules. The head of the spermatozoon acquires the ability to adhere to the zona pellucida, with an increase in net negative charge, after the exposure of sperm to an area of chaperone-laden 'dense bodies', suggesting that the final molecules necessary to complete the zona pellucida-recognition machinery arise in the external epididymal environment. Current evidence suggests that sperm zona pellucida binding proteins link to sialic acid residues within the zona pellucida. However, many other antigens with a demonstrable role in zona binding and fusion are present on the spermatozoa, such as a membrane-bound hyaluronidase (PH20/2B1), fertilin, proacrosin, β1, 4-galactosyltransferase (GalTase) and putative zona ligands sp56 and p95. These are synthesized in the testis as precursors and then activated at some point in the epididymis either through direct biochemical modification, by changing their cellular localization or both. Changes in lectin-binding ability of the sperm plasma membrane during epididymal maturation indicate alterations to the terminal saccharide residues of glycoproteins, and membrane lipids also undergo changes in their physical and chemical composition. In the guinea pig and some marsupials, the acrosome undergoes gross morphological changes at epididymal maturation, while in many eutherian mammals, the nuclear protamines are extensively cross-linked by disulphide bonds imparting rigidity to the sperm head.

In 1952, Chris Austin showed that these maturation events were not sufficient for fertilization and the spermatozoa must undergo further changes in morphology and behaviour in the female reproductive tract to gain fertilization competence. These pre-activation events in mammals are called 'capacitation'. Where capacitation starts and ends in the female reproductive tract varies from species to species. In species where the spermatozoa are deposited in the vagina at coitus, such as humans, cows and sheep, capacitation may begin during passage through the cervical mucus. In animals in which the spermatozoa are deposited directly in the uterus, such as the horse, pig and many rodents, the spermatozoa complete capacitation in the lower segment of the isthmus. Although many structural and biochemical changes may occur in spermatozoa at capacitation, the major events are the removal or alteration of material coating the surface of the epididymal spermatozoa, including 5–10 kDa caltrin, spermine, various glycoproteins, an acrosomal stabilizing protein and various antigens. The phospholipid composition and distribution in the sperm plasma membrane may also change during capacitation as does the efflux and/or lateral movement of cholesterol. Capacitation is regulated by the removal of seminal plasma decapacitating factors and by factors present in the female reproductive fluids by altering the sperm plasma membrane permeability to Ca^{2+}, while it is known that $NaHCO_3$ and serum albumin are also keys components, at least in the mouse. Maturation and capacitation are both influenced by pH_i, which increases during capacitation mainly through a Na^+, Cl^- and HCO_3^- dependent mechanism. Capacitation may be induced in vitro in a variety of animal species and may require from less than one hour in the human and mice, and up to several hours in the rabbit.

The final stage of pre-activation in mammalian spermatozoa is called hyperactivation, characterized by a vigorous flagellar movement, which in the mouse and hamster begins shortly before they are released from the isthmus. Hyperactivity may assist the spermatozoa from breaking free from their bond in the isthmus reservoir and must occur at the right place and time to avoid premature exhaustion before locating or penetrating the oocyte. As in all phases of sperm motility regulation, hyperactivity is dependent on external Ca^{2+} and may be regulated not only by factors in the reproductive tract, such as progesterone, but possibly also by factors released from the oocyte. It is important to stress that populations of spermatozoa are not physiologically homogeneous and that only a small percentage of the cells in the sperm pool in the isthmus (or indeed in vitro) will become capacitated or hyperactivated at any one time. This, we will see in the chapter on fertilization dynamics, may be an adaptation to ensure that the oocyte is exposed over a period of time to fertilization competent spermatozoa.

Chemotaxis

Sperm chemotaxis was first described in a plant, the bracken fern, by Pfeffer in 1884. The sperm attractant was identified in 1958 as α-malic acid. J. Dan in 1950, at the marine laboratory in Woods Hole, demonstrated chemotaxis for the first time in animals using a hydrozoan as a model system. Since then it has been described in Cnidaria, Urochordata, Mollusca, Hydrozoa and Echinodermata. To qualify as a true chemotactic response, the spermatozoon must change swimming direction and/or the waveform pattern of its flagella in response to a gradient of factors released by the oocyte, its coats or the reproductive tract. Frank Lillie over 100 years ago showed that sea urchin sperm behaviour is modulated by factors released from the extracellular coats of the oocyte. Since then, over 70 sperm-activating peptides (SAPs) in the jelly layer from many echinoderms have been identified, mainly from the laboratory of Norio Suzuki in the 1980s. These peptides are composed of between 10–14 amino acids and the most well-studied are speract and resact. Speract is a decapeptide isolated from the sea urchin *Stronglycentrotus*, while resact is a similar peptide isolated from the sea urchin *Arbacia*. SAPs stimulate sperm motility and respiration by a cascade of intracellular signaling events that involve cyclic nucleotides, pH_i and calcium. The receptor for resact on the spermatozoon has been suggested to be guanylyl cyclase. A transient burst in the synthesis of cyclic guanosine monophosphate (cGMP) and cAMP hyperpolarizes the sperm plasma membrane due to K^+ efflux from ion-specific channels, and this leads to the modulation of intracellular calcium through Ca^{2+} gated channels. In the ascidian *Ciona intestinalis*, the sperm chemo attractant is released from the oocyte itself, not the extracellular layers, and appears to be released from the vegetal pole, the preferential site of sperm penetration. The compound in ascidians, that both activates and attracts spermatozoa, has been called SAAF (sperm activating and attracting factor) and appears not to be a protein and does not contain sugar moieties. SAAF causes an increase in K^+ permeability in the sperm plasma membrane resulting in a sustained hyperpolarization and the entry of Ca^{2+} through specific ion channels. An increase in cAMP dependent protein phosphorylation then modulates the sperm movement.

Fertilization in the Octopus is indirect and internal. The male transfers the spermatozoa in a spermatophore into the female oviduct where they are then stored immobile in the spermathecae until ovulation. The ovulated oocytes emit a chemo-attractant peptide of 11 kDa called Octo-SAP, which changes the sperm swimming behaviour in a dose dependent manner by increasing internal Ca^{2+} in the spermatozoa and stimulating membrane protein Tyr phosphorylation.

There is less evidence for chemo attractants in vertebrates, although in the amphibian *Xenopus laevis* a peptide has been identified in the oocyte that appears to attract spermatozoa. Over the last few decades, there has been increasing evidence to indicate that mammalian sperm are able to respond to chemo attractants secreted from the oocyte, the extracellular coats and their surrounding environment. Follicular fluid appears to contain chemotactic factors; however, to date, only progesterone has been identified as a true chemo attractant. It should be noted that, as for capacitation, only a small percentage of spermatozoa are responsive to follicular fluid, and the chemotactic response is transient. Consequently, there is a continuous turnover of responsive spermatozoa in a given population. Although the nature of the progesterone receptor on the sperm is unknown, CatSper, a pH-dependent calcium channel of the sperm flagellum appears to be responsible for the elevation in intracellular Ca^{2+}. Recent evidence has suggested that a protein expressed by the cumulus cells of the oocyte, CRISP 1, stimulates sperm orientation via the CatSper channel, while an odorant receptor gene expressed in the testis may be involved in sperm chemotaxis in humans.

Interaction with the Outer Extracellular Coat

The jelly layer around oocytes, although an essential modulator of sperm progression, is, owing to its nature, difficult to analyse in situ. Jelly organization has been best studied in amphibians, where a fibrillar matrix of high molecular weight glycoproteins is interspersed with globular proteins of lower molecular weight. At the time of deposition, the amphibian jelly layer is also extremely rich in ions, containing 70 mM Na^+, 30 mM K^+, 6 mM Ca^{2+} and 7 mM Mg^{2+}. Leaching these ions, in particular calcium and the low molecular glycoproteins, in hypotonic medium, leads to a marked yet reversible decrease in fertilizability. Considering that in anurans spermatozoa enter the oocyte in the animal hemisphere, it is probable that the jelly layer differs from antipode to antipode. The oocyte jelly layer of the frog *Discoglossus pictus*, as in other amphibians, contains a sperm motility–initiating substance (SMIS). When the sperm's apical rod (the apex of the head) makes contact with the jelly layer, motility is initiated, but it lasts only 14 seconds. In this time, the spermatozoa converge into the funnel-like dimple area (Figure 4.3) and make contact with the oolemma.

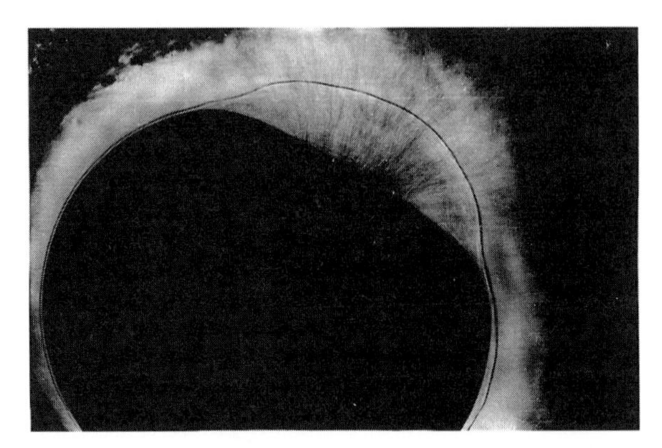

Figure 4.3 Spermatozoa in the dimple area of the painted frog *Discoglossus pictus* (courtesy of Professor Chiara Campanella).

The jelly layer in echinoderms (see Figures 3.1 and 3.2) is composed mainly of high molecular weight glycoconjugates rich in fucose-sulphate; however, little is known about its structure in situ. The pH in the jelly layer of some species is lower (6.3 to 7.6) than that of the surrounding seawater (7.8 to 8.2), which may play a role both in maintaining the metabolic repression of the oocyte and in regulating sperm–oocyte interactions. It needs to be clarified whether the jelly layer is subdivided into micro-environments, both radially and topographically, that differ with regard to sperm recognition and progression. In fact, the schools of Boveri at the beginning of the twentieth century and Runnstrom in the 1960s considered the jelly coat to be a barrier to spermatozoa with a preferential entry site, the jelly canal at the animal pole, analogous to the micropyle in fish and insect oocytes. Berndt Hagstrom in the 1960s, showed that even at relatively high sperm to oocyte ratios, 90 per cent of spermatozoa are unable to penetrate the jelly layer in sea urchin oocytes and remain immobilized at various depths within the jelly. Those that pass through must arrive in a physiological condition that both promotes binding and subsequently penetration of the vitelline membrane.

The extracellular matrix in mammalian oocytes, called the cumulus oophorus, is a complex of cells and hyaluronic acid (Figure 4.4), secreted by the cumulus cells at meiotic resumption prior to ovulation. Hyaluronic acid, the major component, is a large

(a)

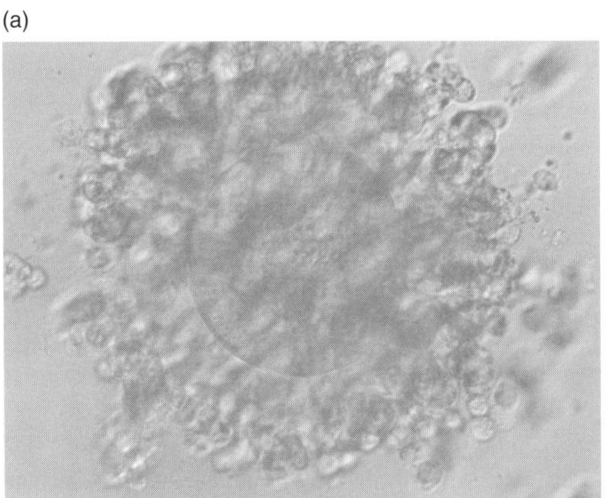

(b)

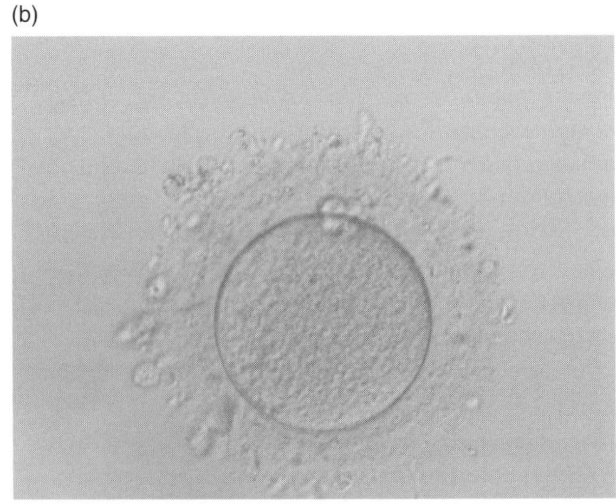

Figure 4.4 **(a)** A human oocyte shortly after aspiration from a follicle showing the extensive and thick cumulus composed of cells and a hyaluronic acid matrix. **(b)** The same oocyte after the cumulus has been removed showing the polar body. (A black-and-white version of this figure will appear in some formats. For the colour version, please refer to the plate section.)

polymer of alternating N-acetylglucosamine and glucoronic acid residues with an average molecular mass of 2500 kDa. There are also several protein components in the extracellular matrix which may stabilize it, such as fibronectin, laminin and tenascin-C. Laminin is uniformly distributed throughout the matrix, whereas fibronectin is only present around the corona radiate. In most eutherian mammals, the cumulus oophorus remains around the ovulated oocyte until fertilization is complete, meaning that the spermatozoa must interact with and navigate through this substrate. In contrast, in marsupials and monotremes the cumulus mass is shed at ovulation and does not present a barrier to the spermatozoa.

Whether or not the cumulus oophorus induces the acrosome reaction in mammals is a matter for conjecture; however, it does have other clear functions, such as the transport of the oocyte into the oviduct, and it is also the source of a variety of soluble factors and hormones that affect the behaviour of both oocyte and spermatozoa. These factors may act before ovulation to induce the resumption of meiosis and after ovulation to trigger sperm hyperactivity. The high visco-elasticity of the hyaluronic acid surrounding the cumulus cells presents a barrier for sperm penetration. Sperm are presumably endowed with an enzyme that can break down the surrounding cumulus which should be located externally since acrosome reacted spermatozoa are unable to progress through the cumulus layer.

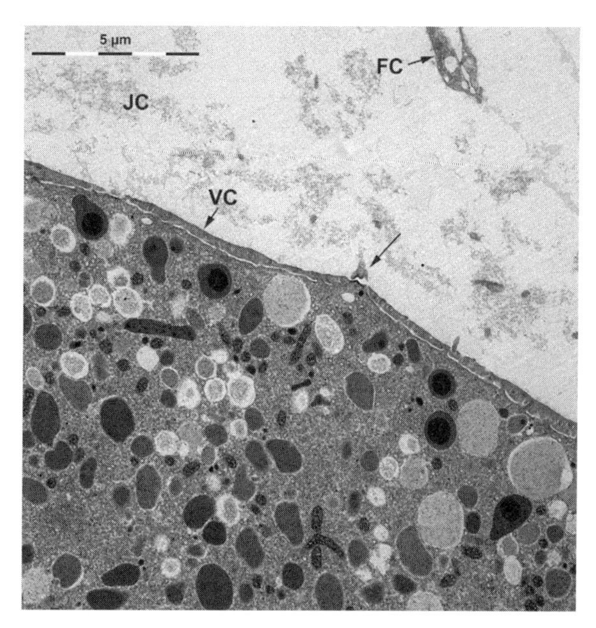

Figure 4.5 The surface of an immature starfish oocyte at the transmission electron microscope showing the vitelline coat (vc), the jelly coat (JC) and follicle cells (FC). The arrow shows a protrusion of the follicle cell making contact with the oocyte plasma membrane through a hole in the vitelline coat (courtesy of Dr Luigia Santella, Stazione Zoologica).

Interaction with the Inner Extracellular Coat

The specialized inner layer of the extracellular coats of oocytes is called the zona pellucida in mammals, the vitelline envelope in echinoderms and amphibians and the chorion in ascidians. This inner oocyte coat is composed of proteins and carbohydrates in the form of glycoprotein units which are probably stabilized by disulphide bonds. The principle carbohydrate appears to be fucose, and the glycoprotein units are synthesized by the oocyte itself. The form of the vitelline coat varies greatly from species to species. For example, in the sea urchin, the coat is very thin and adheres tightly to the oocyte surface, following the contours of the surface microvilli; whereas in the starfish, it is much thicker and is perforated by the microvilli (Figure 4.5). In mammals and ascidians, the vitelline coat is relatively thick and is separated from the oocyte surface by a space called the perivitelline space. In the ascidians this space, is filled with a further layer of cells called the test cells (see Figure 3.4). The zona pellucida in the mouse is a compact, highly organized matrix approximately 7 µm thick with a lacey appearance. The long, interconnected fibrils are held together by non-covalent bonds. A section cut through the human zona pellucida and observed at the scanning electron microscope can be seen in Figure 4.6. Note also the dense organization of the external cumulus cells. The successful spermatozoon must navigate through these cells and penetrate the zona pellucida in order to make contact with the oocyte surface.

The glycoproteins that make up the zona pellucida (ZP) are highly conserved among mammals. In the mouse, the ZP is made up of three glycoproteins, ZP1, ZP2 and ZP3, with molecular masses of 200, 120 and 83 kDa respectively. All three ZP proteins are glycosylated with aspargine-(N) and serine/threonine-(O) linked oligosaccharides. A structural element, the ZP domain (ZPD), composed of 260 amino acids, is found in all ZP proteins as well as many other proteins with different functions, such as receptors, and in intracellular signaling, differentiation and morphogenesis. There are 10 ZPD proteins in nematodes,

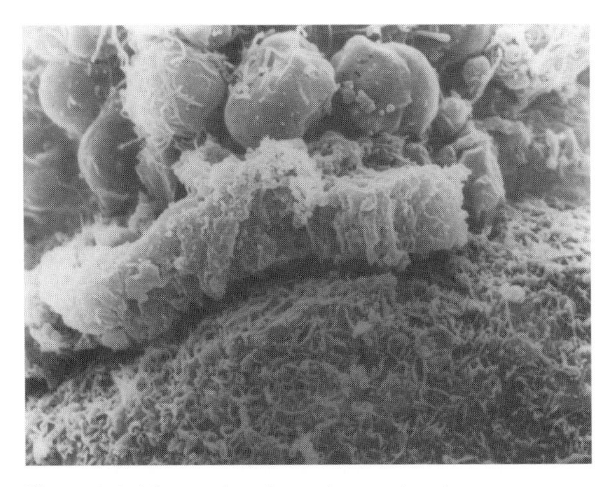

Figure 4.6 A fractured section at the scanning electron microscope through the zona pellucida of a human M11 oocyte showing the surface microvilli of the plasma membrane and the extensive organization of cumulus cells. The spermatozoon has to navigate through these cells to reach the oocyte.

more than 20 in flies and over 100 in birds. Human zona pellucida contains an additional glycoprotein, ZP4. Synthesis of these glycoproteins is temporally regulated during oogenesis with the ZP genes transcribed exclusively by the oocytes and/or follicle cells in mammals and amphibians. ZP2 is expressed at low levels in resting oocytes, but ZP1 and ZP3 are only expressed by growing oocytes. In fish and birds, the ZP proteins are synthesized in the ovary and/or the liver. ZP3 is an 83 kDa glycoprotein with three or four N-linked oligosaccharide chains and a number of O-linked oligosaccharide chains and appears to be the primary adhesion molecule. It is thought that the bioactive component within ZP3 is related to its carbohydrate composition with the terminal sugar residue being either a terminal alpha-linked galactose or a terminal N-acetylglucosamine. Other studies have suggested that ZP1 may also be involved in primary adhesion events at least in the rabbit and the human. ZP3 and ZP2 are the two major sub-units of the zona pellucida and it has been suggested that the N-terminal region of ZP2 regulates recognition in the mouse and the human. When sperm bind to ZP3 they undergo the acrosome reaction.

The vitelline envelope in fish, reptiles, amphibians and birds also consist of a limited number of proteins all closely related to ZP1–4. For example, the vitelline envelope in the oocyte of the amphibian *Xenopus laevis* is composed of three glycoproteins, homologous to the mouse ZP at 30–40 per cent amino acid identity. It seems that oocyte coat proteins derive from a common ancestral gene, possibly ZP3, which duplicated several times hundreds of millions of years ago to give rise to three to four genes in fish, four to five genes in amphibians, six genes in birds and three to four genes in mammals.

In the sea urchin, the acrosomal protein bindin, exposed following acrosomal exocytosis and characterized mainly by Victor Vacquier's group in California, mediates adhesion of the spermatozoon to the vitelline envelope. Bindin sequences are specific to the species and have a central domain of about 60 amino acids, which has been conserved during more than 150 million years of evolution. The species specificity of bindins in sea urchins is due to the number and location of short repeating amino acid sequences that flank the central conserved domain. Sea urchin bindins are unique proteins and are not related to any other proteins found in the animal kingdom.

In ascidians, the primary binding of the spermatozoon to the chorion is thought in part to be a substrate enzyme-like reaction between a sperm located α-L-fucosidase and the complimentary L-fucosyl residues of glycoproteins on the chorion. Other molecules involved in this initial adhesion in ascidians are trypsin-like acrosin and spermosin protease. Ascidian sperm have a chymotrypsin-like protease on their surface that appears to digest a passage through the chorion.

The vitelline envelope of abalone oocytes, a marine mollusc, also studied by the group of Victor Vacquier in California, is about 0.6 μm thick and is composed of glycoprotein filaments. Sperm bind to the vitelline envelope by the plasma membrane of the sperm over the acrosomal vesicle. The vesicle opens by exocytosis revealing a non-glcosylated 16 kD lysin and a 18 kD protein. The lysin dissolves a 3 μm hole in the vitelline envelope by a non-enzymatic way, probably by disrupting the hydrogen bonds that hold the vitelline envelope filaments together. The 18 kD protein coats the surface of the acrosomal process as it extends to a length of 7 μm and is thought to be a fusogen for the successive fusion with the plasma membrane of the oocyte. The receptor for lysin is called VERL and this glycoprotein comprises about 30 per cent of the vitelline envelope. The functional N-terminal sperm-binding region of VERL consists of 22sequence repeats spaced by linkers and is structurally homologous to the functional region of the mammalian ZP2.Therefore, despite being separated by 0.6 billion years, molluscs and humans use a

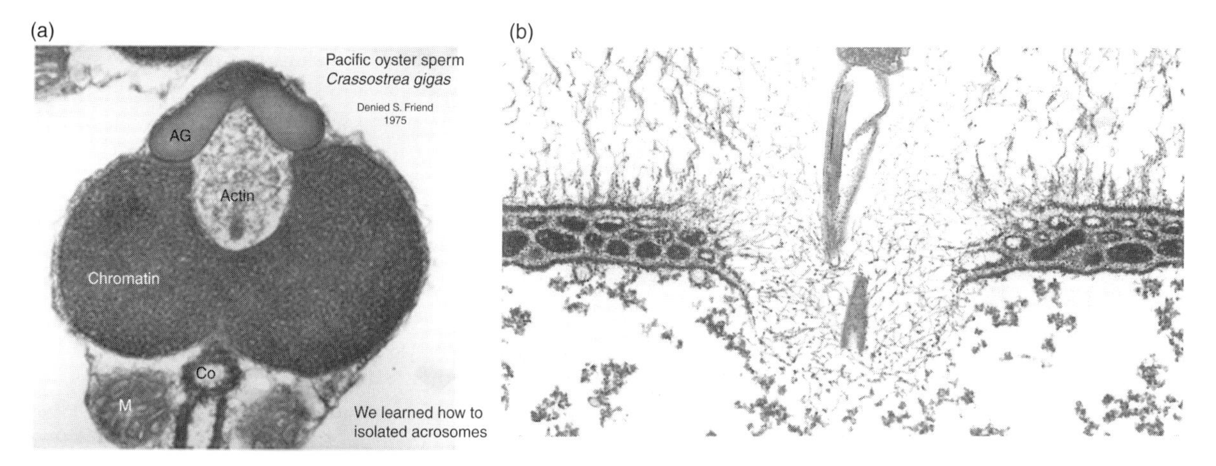

Figure 4.7 A transmission electron micrograph through the head region of an oyster spermatozoon showing the ring-shaped acrosome (courtesy of Professor Victor Vacquier).

common C-terminal ZP module as a building block for the assembly of their oocyte coats showing the similarity between the invertebrate vitelline envelope and the zona pellucida.

In another mollusc, the oyster, the spermatozoa have ring-shaped acrosome vesicles that exocytose to expose an insoluble protein that bind the spermatozoa to microvilli on the egg vitelline envelope (Figure 4.7). The insoluble acrosomal rings contain oyster bindin, a protein that agglutinates unfertilized eggs. The putative functional unit of bindin is a fucose lectin (F-Lectin). Sperm from the pacific oyster contain bindin proteins of 35, 48, 63, 75 and 88 kDa; all products of a single gene.

The Acrosome Reaction in Invertebrates and Vertebrates

The interaction of the spermatozoon with the oocyte involves many overlapping events from the initiation of motility, through primary and secondary binding to gamete fusion. Clear start and end points are difficult to establish. The site of the acrosome reaction is no exception, particularly in mammals.

Perhaps the clearest example of the acrosome reaction and where and when it occurs is the starfish oocyte, since it is clearly visible at the light microscope. Hermann Fol in Naples in 1890, noted the acrosomal tubule extending through the jelly layer and photographic documentation was presented nearly 90 years later from my group in the same laboratory (Figure 4.8). Upon encounter with the

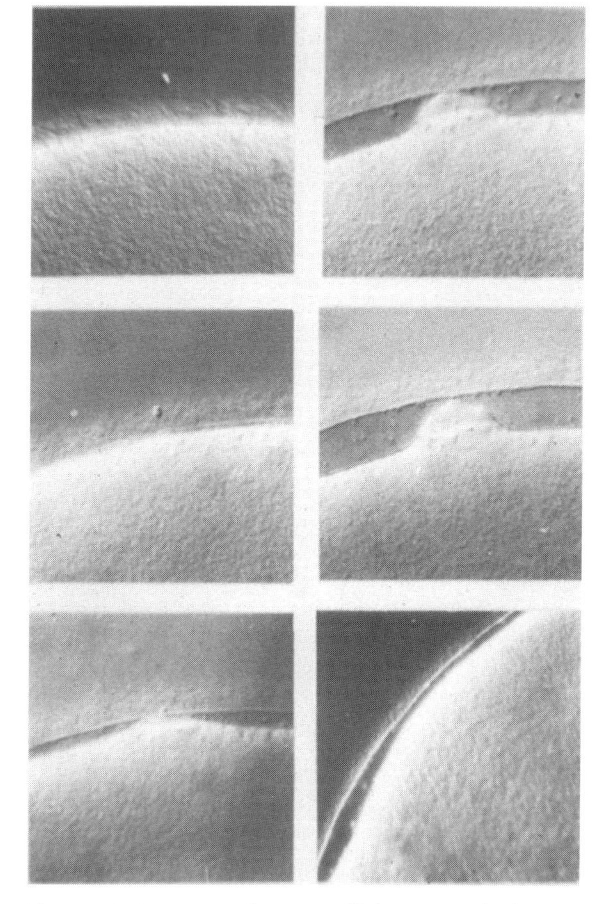

Figure 4.8 A sequence of transmitted light micrographs showing the acrosome reaction in the starfish *Astropecten aurantiacus*. In the top left frame, the acrosomal tubule can be seen extending through the jelly layer to the oocyte surface.

jelly layer intracellular Ca^{2+} and pH_i in the spermatozoon increase, triggering exocytosis of the acrosomal vescicle and the acrosome reaction. The acrosomal tubule, which is about 25 μm long, is emitted obliquely to the axis of the head and flagellum piercing the jelly layer and the underlying vitelline membrane to fuse with the oocyte plasma membrane (Figure 4.9). The starfish jelly coat contains many organic compounds and is mainly composed of glycoproteins. The laboratory of Motonori Hoshi in Japan demonstrated that three of these jelly compounds act together to induce the acrosome reaction. They are ARIS (acrosome reaction inducing substance), which is a highly sulphated proteoglycan-like substance; Co-ARIS, a group of sulphated steroid saponins, and a group of sperm-activating peptides called Asterosap. Individually these compounds do not trigger a complete acrosome reaction, for example, ARIS and Co-ARIS alone will increase Ca^{2+} uptake by the spermatozoa, while Asterosap alone will only increase the intracellular pH_i. ARIS proteins are found in all invertebrates from ctenophores to cephalochordates. The only exception is the sea urchin, which seems to have abandoned this protein in favour of a different mechanism.

In the sea urchins, a fucose sulphate polymer (FSP) in the oocyte jelly is responsible for triggering the acrosome reaction. Here acrosomal exocytosis leads to the release of the acrosomal contents, which include many enzymes, extension of the acrosomal tubule and exposure of bindin, a 30,000 molecular weight polypeptide characterized in Victor Vacquier's laboratory in California in the 1980s that is required for adhesion of the spermatozoon to the vitelline membrane. Na^+ and Ca^{2+} entry and K^+ and H^+ efflux in the spermatozoon are activated within seconds of exposure to FSP as a consequence there is a change in resting membrane potential, intracellular Ca^{2+} and pH_i. Evidence for two different Ca^{2+} channels and their synergistic action in inducing the acrosome reaction has been reported, while the Na^+/H^+ exchange and rise in intracellular pH is induced by mobilization of K^+ channels, causing a fast and transient hyperpolarization followed by a Ca^{2+}-mediated depolarization. Cl^- selective anion channels also identified in the sea urchin sperm plasma membrane may play a role in the acrosome reaction. cAMP, inositol-1,4,5 triphosphate ($InsP_3$), phospholipase D and nitric oxide synthase all increase at the acrosome reaction. The receptor on the spermatozoon for the egg jelly factor has been cloned and is thought to be a sperm plasma membrane protein of 210 kDa called REJ. This receptor has an epidermal growth factor module, two C-type lectin carbohydrate-recognition modules and a 700-residue module that has extensive homology with the human kidney disease protein polycystin-1.

The hemichordate Saccoglossus, was one of the first animals to be used as a model to describe the acrosome reaction. The membrane-bound acrosomal granule contains lytic agents such as proteases,

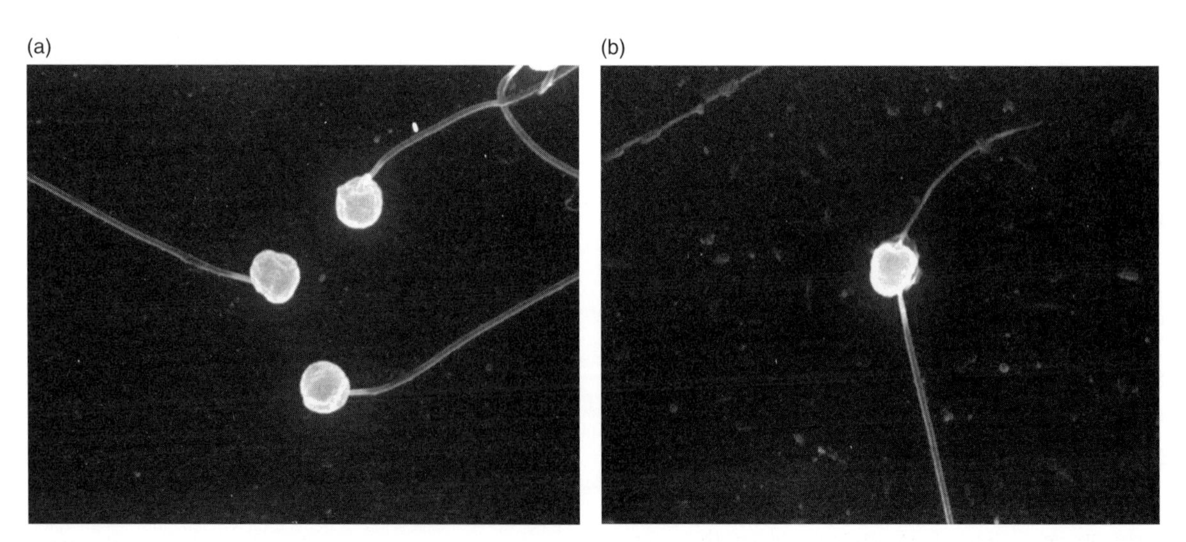

Figure 4.9 **(a)** Unreacted spermatozoa of the starfish showing the typical round shape and **(b)** the acrosome reacted spermatozoon showing the oblique angle of the acrosomal tubule and the dumbbell shaped head (courtesy of Dr Keiichiro Kyozuka).

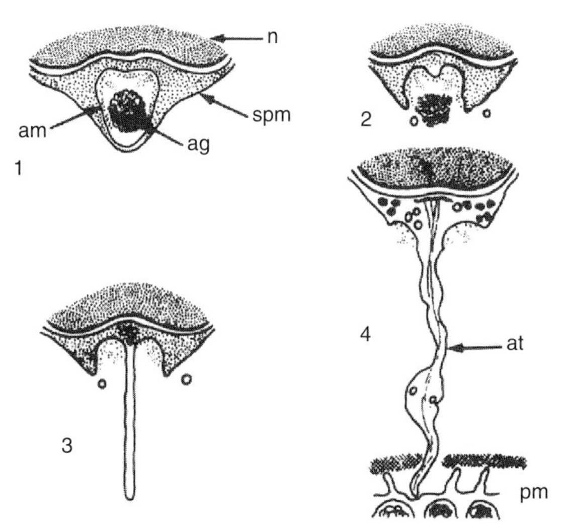

Figure 4.10 The acrosome reaction in the hemichordate Saccoglossus. Fusion of the sperm plasma membrane with the acrosomal membrane releases the acrosomal granule. The extending acrosomal tubule then makes contact with the plasma membrane of the oocyte. This reaction takes seven to nine seconds (from Dale, 1983).

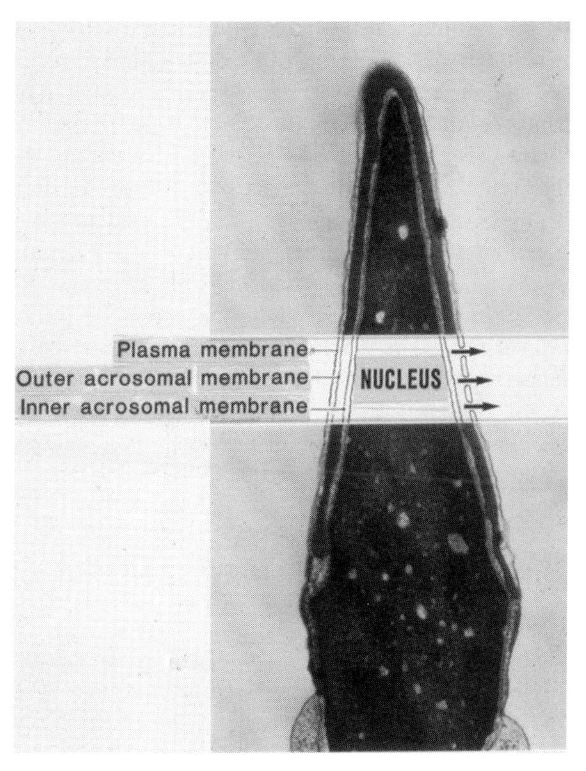

Figure 4.11 A section through a human spermatozoon at the transmission electron microscope showing the layout of membranes forming the acrosomal cap.

sulphatases and glycosidases, and is bound within the plasma membrane. When the sperm attaches to the vitelline coat, the permeability of the sperm plasma membrane is altered, causing a transient change in the concentration of intracellular ions which triggers the acrosome reaction. There are four stages in the AR of Saccoglossus. The acrosome membrane fuses with the sperm plasma membrane at the tip of the acrosome, causing acrosomal granule breakdown. The released lysins either dissolve a pathway through the vitelline coat or alter it to allow the penetration of the acrosomal tubule. This structure is formed by the polymerization of actin extending the inner acrosomal membrane until it contacts the oocyte plasma membrane and the two membranes fuse (Figure 4.10). The first two steps are thought to occur within a second, while all the stages up to and including fusion may take seven to nine seconds in this hemichordate.

The structure of the sperm acrosome in the human is shown in Figure 4.11 and is quite representative of most mammals. Essentially, the acrosome, containing enzymes and many other compounds, is bordered by an inner acrosomal membrane that overlies the sperm nucleus and an outer acrosomal membrane that lies directly below the sperm plasma membrane. Acrosomal exocytosis is calcium dependent and results from multiple fusion sites between the sperm plasma membrane and the outer acrosomal membrane. The molecular mechanisms of plasma membrane and outer acrosomal membrane recognition, docking and adhesion leading to fusion are common to other secretory events and are most probably primed during the process of capacitation. Two classes of proteins involved in protein-mediated membrane fusion are SNARES (soluble NSF-attachment protein receptors) and the Rab proteins, which are monomeric GTP-binding proteins. The patches of ruptured membrane allow the efflux of the digestive enzymes and exposure of the highly fusogenic inner acrosomal membrane.

In mammals there are contradictory views as to where the acrosome reaction takes place. Confusion has probably arisen from interpretation of data from in vitro fertilization studies rather than in vivo studies and the fact that the sperm population is extremely heterogeneous with a continuous but transient replacement of capacitated sperm in the sperm pool. Certainly, solubilized ZP or purified ZP3 will induce the acrosome reaction, however hamster and rabbit sperm seem to undergo the acrosome reaction as they progress through the cumulus. Progesterone, secreted by the cumulus cells, will also initiate the acrosome reaction in mammals suggesting the reaction starts in the cumulus oophorus. After the spermatozoon binds to its receptor, a GTP-binding protein and phospholipase C are activated and lead to an elevation in intracellular Ca^{2+} through T and L-type calcium channels and CatSper channels. The membrane potential of the spermatozoon hyperpolarizes due to increased activity of K^+ channels, and a decrease in Na^+ channel activity also causes an increase in intracellular Ca^{2+}. The mammalian sperm head also contains several type of Cl^- channel, while the cytoplasm is rich in Cl^- ions. Intracellular pH_i also increases following the G-protein-dependent pathway, apparently in conjunction with an increase in cAMP production.

The Chorion and Micropyle in Fish, Insects and Squid

Oocytes of teleost fish only allow sperm entry at a single site, a narrow canal in the chorion called the micropyle. In some fish (e.g. the flounder and the herring), the micropyle is narrow sometimes with a shallow depression around the outer opening (Figure 4.12), while in others (e.g. salmon and cod) it has a funnel-like conical opening. In all the above species, a glycoprotein on the chorion surface around the micropyle directs spermatozoa into the canal. This substance, called the micropylar sperm attractant or MISA, increases the probability of fertilization. The mechanism underlying sperm–MISA interactions has yet to be determined, but at least in herring the involvement of Ca^{2+} and K^+ channel proteins, as well as CatSper and adenylyl cyclase, is very likely. Zebrafish and goldfish oocytes, in contrast, do not contain MISA. Here the chorion around the micropyle is deep and may have radially or spirally arranged grooves at the mouth (Figure 4.13). In fish, there is no indication that spermatozoa are attracted by oocytes from a distance, but once near the outer opening of the micropyle, they exhibit directed movement towards it, a chemotactic-like response. The Pacific herring *Clupea pallasi* is unusual in that, in addition to MISA, the oocyte has two other sperm activating factors. A factor HSAP (herring sperm activating protein) that is released from the oocyte surface and serves to increase the motility of the spermatozoa in the general surroundings of the oocyte and SMIF (sperm motility initiating factor) that is tightly bound to the chorion. SMIF is a 105 kD water insoluble glycoprotein. It is not sure if SMIF and MISA in the herring are the same molecule. B. Afzelius showed in 1978 that the head of spermatozoa in many teleosts are small and round (diameter of 2–4 μm), and they do not have acrosomes (see Figure 3.15), while in sturgeons and paddlefish the sperm heads are elongate (10 μm by 2 μm wide) and acrosomes are present. Perhaps the presence of the acrosome is related to the oocyte structure, since in the former group the chorion has a single micropyle and in the latter group the oocyte has multiple micropyles.

Oocytes of some insects, including *Drosophila*, also have distinctive glycoprotein micropyle caps, which also may prove to assist sperm entry (Figure 4.14).

Fusion of the Plasma Membrane of the Spermatozoon with the Oocyte

The plasma membrane is not, as once thought, a random ocean of lipids. The lipids are distributed asymmetrically and organized into nanometer-sized subdomains, ranging from 10 to 200 nm, called lipid rafts with specialized regions of high cholesterol, sphingomyelin, gangliosides and enriched in phospholipids with saturated fatty acyl chains. A major raft constituent is Caveolin, which serves as a scaffolding to embed and inactivate many proteins and enzymes. Lipid rafts are less fluid than the rest of the plasma membrane and display lateral movement in response to physiological stimuli. Since they contain a variety of membrane proteins they are thought to be platforms for membrane trafficking, signal transduction and cell adhesion events such as viral entry. Individual rafts, although small, are thought to constitute a large fraction of the plasma membrane, comprising up to 20 per cent of the surface area of somatic cells. In mouse, sea urchins and amphibians, fertilization may be inhibited by raft disruption using methyl-beta-cyclodextrin (MBCD), which disperse important raft proteins such as CD9 and inhibits Src activation and the completion of meiosis. MBCD in mouse oocytes disrupts both planar and caveolar rafts, which are thought to be the sites of mammalian sperm–egg binding and fusion. Phosphatidic acid (PA) is concentrated in rafts and may play a role in membrane events by raft stabilization. Other lipids may also affect rafts and membrane fusion, for example production of ceramide during fertilization may lead to clustered rafts and an increase in raft diameter. Cortical microfilaments are important in

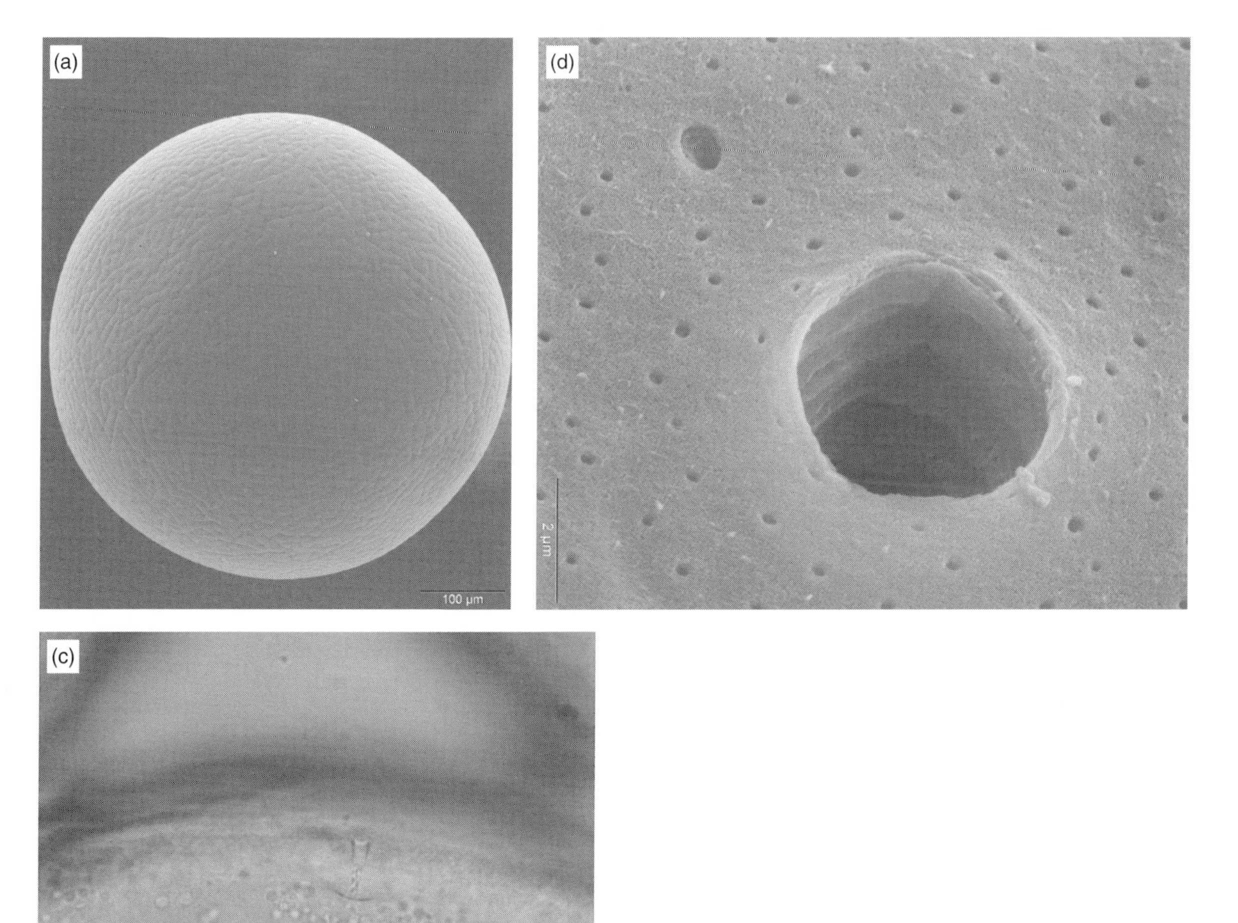

Figure 4.12 **(a)** The micropyle in the oocyte of the black flounder, **(b)** at higher magnification and **(c)** at the light microscope with several spermatozoa in the canal. The canal is 6 μm wide (courtesy of Professor Ryuzo Yanagimachi). (A black-and-white version of this figure will appear in some formats. For the colour version, please refer to the plate section.)

raft biology and PA is a major regulator of cytoskeletal fibres.

Phosphatidic acid can activate Src tyrosine kinase, or phospholipase C during fertilization in *Xenopus* oocytes, leading to an increase in intracellular calcium. In fact, several lipids seem to have roles in membrane fusion events at fertilization, including the acrosome reaction, gamete fusion and cortical exocytosis, regulating receptors and releasing intracellular calcium. Lipases, such as phospholipase D, C and A2, sphingomyelinase, lipin 1 and autotaxin are involved in generating lipid second messengers at fertilization.

In the human oocyte, lipid raft micro-domains are enriched in the ganglioside GM1 and the tetraspanin protein CD9. GM1 is involved in a variety of processes such as virus docking, signal transduction and protein binding, while CD9 seems to be the most important membrane component involved in sperm penetration in mammals. In the human, it appears that sperm penetration into the oocyte is dependent on the density and organization of GM1 micro-domains at the site where the sperm arrives, which can be considered docking sites. The sperm bind at these GM1 micro-domains but do not penetrate. The lipid rafts with CD9, distinct to those with GM1, are a likely candidate for stable binding. The stability of these plasma membrane rafts and GM1 organization depends on the underlying mitochondrial activity and efficiency.

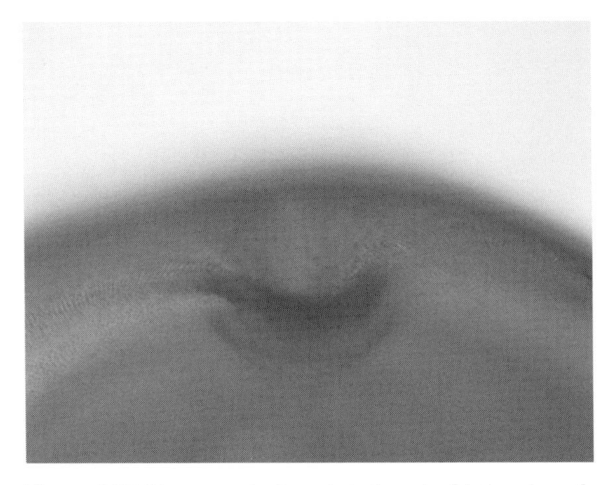

Figure 4.13 The grooved micropyle in the zebrafish (courtesy of Professor Ryuzo Yanagimachi). (A black-and-white version of this figure will appear in some formats. For the colour version, please refer to the plate section.)

Over the past few decades there have been many attempts to identify the molecules on the oocyte surface and on the sperm acrosomal membrane responsible for gamete fusion. Izumo 1(named after a Japanese marriage shrine), is a transmembrane protein in mouse spermatozoa that interacts with Juno (after the Roman goddess of fertility and marriage) a receptor found on the surface of the unfertilized oocyte. Juno is a member of the folate-receptor family known as Folr4 and is a glycophosphatidylinositol (GPI) anchored receptor. To date, Juno has been observed to be a receptor on the oocytes of mice, opossums, pigs and humans. After fertilization, Juno is rapidly lost from the oocyte surface. It appears that the Juno–Izumo1 interaction is a necessary adhesion step for gametes rather than a membrane fusion event and that fusion requires other membrane proteins perhaps such as EEF-1 and myomaker. Local clustering of Juno on the oocytes membrane to increase the strength of sperm binding may be organized by the tetraspanin CD9. This latter protein in its own right

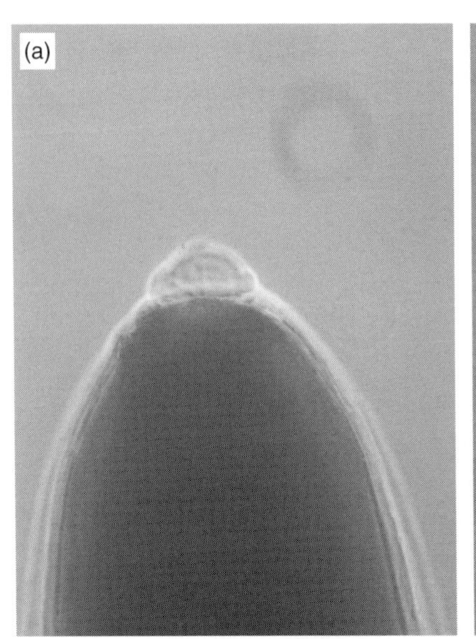

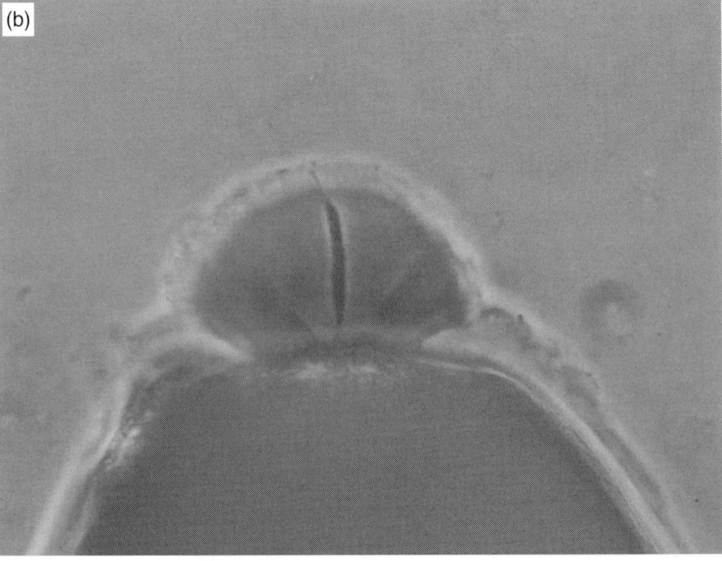

Figure 4.14 **(a)** The jelly cap and the micropyle in the oocyte of the housefly *Mus Domestica* and **(b)** at higher magnification showing a spermatozoon in the micropyle. The jelly cap is some 46 µm wide (courtesy of Professor Ryuzo Yanagimachi). (A black-and-white version of this figure will appear in some formats. For the colour version, please refer to the plate section.)

has a clear role in sperm–oocyte fusion, however the corresponding sperm ligand is unknown. Members of the CRISP family (cysteine rich secretory proteins) are also known to participate in the process of capacitation and mammalian sperm–zona pellucida interactions. The lipid rafts may concentrate signalling proteins at the site of gamete interaction to promote adhesion and fusion. Plasma membrane fusion of the two gametes leading to cell–cell continuity is the last step in the interaction of gametes that leads to oocyte activation and the formation of the zygote. Although the mechanism is still not clear, lipid rafts and associated transmembrane proteins Izumo 1, Juno and CD9 play a fundamental role.

Chapter 5

Oocyte Activation

Oocytes are programmed to interact with the fertilizing spermatozoon at a particular moment in time. Once activated, the emphasis passes from that of a quiescent cell concerned with enticing its male counterpart to a metabolically dynamic zygote that must progress through early development with minimum hindrance or interference from the environment. The first signs of oocyte activation are changes in the ion permeability of the plasma membrane, which allows the entry of external Ca^{2+} and other ions. This is followed by the release of intracellular Ca^{2+} from internal stores, such as the endoplasmic reticulum. While the former maybe localized to the site of sperm entry, the latter starts at the entry site and spreads in an autocatalytic wave to the antipode. Simultaneously, there is a massive reorganization of the cell cortex involving cytoskeletal elements and a cascade of activity of cell cycle kinases. The motors for oocyte activation are MPF (maturation promoting factor) and CSF (cytostatic factor), both discovered by Y. Masui in the 1970s. The regulation of the cell cycle in somatic cells was found at a later date to follow the same principles and use the same molecules. Most activation events are propagative and have precise temporal and spatial characteristics that are predetermined and independent of sperm concentration. Many of the activation events are not specific to gametes, but are biochemical and physiological processes common to all cells, organized in a different way and involving hundreds or thousands of biochemical and physiological pathways. The complex process of activation does not lend itself to traditional molecular biological studies, and gene disruption experiments should be interpreted with caution.

How Does the Spermatozoon Activate the Oocyte

The idea that spermatozoa activate oocytes by releasing a soluble cytolytic factor dates back to Jacques Loeb in 1912 at the beginning of the era of chemical embryology. Frank Lillie, shortly afterwards in 1922, favoured the idea of an externally located receptor-ligand mechanism of oocyte activation that persisted until the mid-1970s. Measurement of membrane noise in the sea urchin oocyte showed that the spermatozoon gated several hundred nonspecific ion channels in the oocyte plasma membrane at fertilization and, by analogy with known events at the nerve-muscle ending, led to the suggestion that this must be due to a soluble factor that is released from the spermatozoon into the cytoplasm (Figure 5.1). However, it was not until 1985 that microinjection experiments gave direct support to the hypothesis of a soluble sperm factor (Figure 5.2). In the 1980s, a G-protein hypothesis was again popular among many scientists, where the spermatozoon acted as an 'honorary hormone' sitting on an externally located oocyte receptor that, via a G-protein, transduced this signal to the oocyte interior, triggering the formation of IP_3 and activating the oocyte. The debate continued through the 1990s, when soluble factors were extracted from mammalian sperm that triggered calcium oscillations in homologous oocytes and the clinical technique of microinjecting spermatozoa into human oocytes showed that the plasma membrane events at fertilization could be bypassed. At the beginning of this century, phospholipase C zeta 1(PLCζ) was suggested to be the soluble sperm factor in mammalian spermatozoa, since recombinant PLCζ RNA or protein can trigger calcium oscillations when injected into oocytes, and depletion of PLCζ from sperm extracts eliminates their ability to trigger calcium oscillations. Recently, using gene knockout technology, it was shown that sperm from $Plcz1^{-/-}$ males failed to trigger calcium oscillations after being microinjected into oocytes. However, some oocytes did develop in vitro, and $Plcz1^{-/-}$ males are only sub-fertile. They are able to father pups after copulation in vivo, albeit producing slightly smaller litter sizes, suggesting that other pathways are available for oocyte activation.

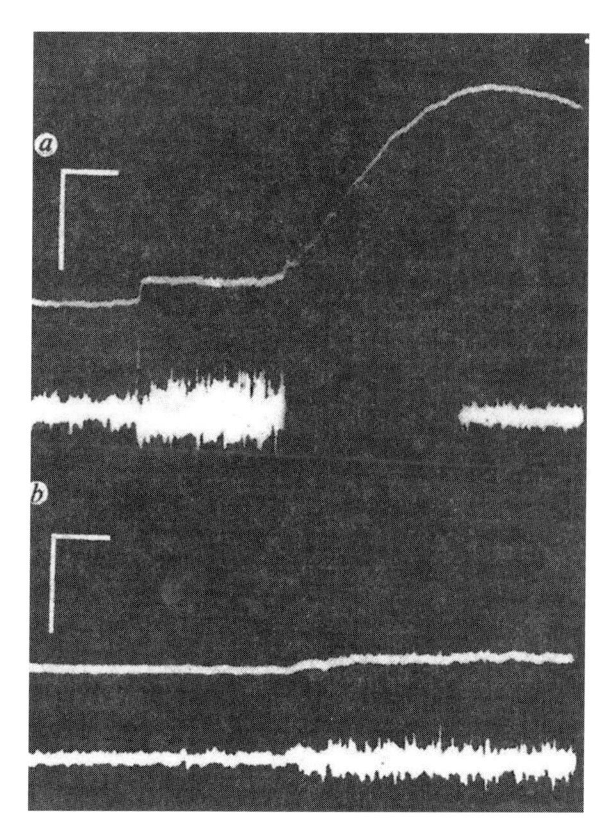

Figure 5.1 Changes in voltage in a sea urchin oocyte at fertilization. The lower traces show membrane noise. The fertilizing spermatozoon triggers a step-like change in voltage by activating non-ion specific channels in the oocyte plasma membrane of 33 pS. The delay between the two electrical events is the latent period. This was the first suggestion of a soluble sperm-borne activating factor at fertilization. The vertical bar is 12 mV, the horizontal bar is 5 seconds (from Dale et al. 1978).

To date, there are many candidates both for soluble sperm factors that are released into the oocyte cytoplasm and 'externally located receptors' found on the surface of the oocyte plasma membrane that function through transmembrane signalling. For example in mammals, although PLCζis a favoured sperm factor, a truncated c-kit tyrosine kinase and a post-acrosomal sheath, WW domain binding protein (PAWP) remain in contention.

In invertebrates, simpler molecules have been suggested to be the sperm-borne activator, such as calcium ions, inositol triphosphate, cyclic-guanosine monophosphate, adenosine diphosphate ribose, NAADP and nitric oxide, which are all second messengers known to be present in spermatozoa. In fact, it has been shown that a spermatozoon of the ascidian *Ciona intestinalis* contains 1–3 x 10^{-19} moles of IP_3, while a human spermatozoon contains 6 x 10^{-19} moles of IP_3, which corresponds to intracellular concentrations of 50–200 μm; much higher than the estimated 6 μm found in sea urchin oocytes. It is interesting that ascidian spermatozoa maybe pre-treated with lithium or phorbol ester, which both affect the phosphoinositide (PI) cycle of the sperm, and when these sperm are used to inseminate oocytes, the peak amplitude and slope of the fertilization current is dramatically increased (Figure 5.3), showing that the kinetics of an oocyte activation event can be altered by changing the physiological status of the spermatozoon. In cross-fertilization experiments, sperm from the oyster *Ostrea edulis* are able to induce the fertilization current in ascidian oocytes and therefore can gate the same ion channels that are gated by homologous spermatozoa; however, these heterologous sperm cannot induce the surface contraction. The oyster sperm enters the ascidian oocyte at the same predetermined site at the vegetal pole as used by homologous spermatozoa. Gating of the fertilization channels by a spermatozoon from a different phylum implies a universal mechanism for oocyte activation, at least in part. Injection of soluble extracts from human sperm into ascidian oocytes triggers the initial waves of calcium in these oocytes but not later events such as the second set of calcium oscillations or second polar body extrusion, suggesting that soluble sperm factor contains several components that trigger multiple independent events. After sperm entry in ascidians, the sperm aster remains at the vegetal pole for the duration of meiosis 1 and then migrates to the equator of the oocyte in meiosis 2. The phase-3 calcium oscillations originate from the male aster in the equatorial region, which behaves as a pacemaker that also controls the speed of the calcium waves. The male centrosome develops into a large tubulin-containing complex that divides to form the mitotic apparatus, while the female centrosome remains small and degrades. Interestingly, a cytoplasmic factor must inhibit the female centrosomal complex, whereas the sperm centrosomal complex is protected from this factor.

In amphibians, there appear to be both soluble sperm factors and externally located ligand-messaging systems involved in oocyte activation. For example, in *Bufo arenarum,* two fractions have been extracted from spermatozoa: a 24 kD fraction that induced oocyte activation when injected into the oocyte and a 36 kDa protein that induced activation when applied to the

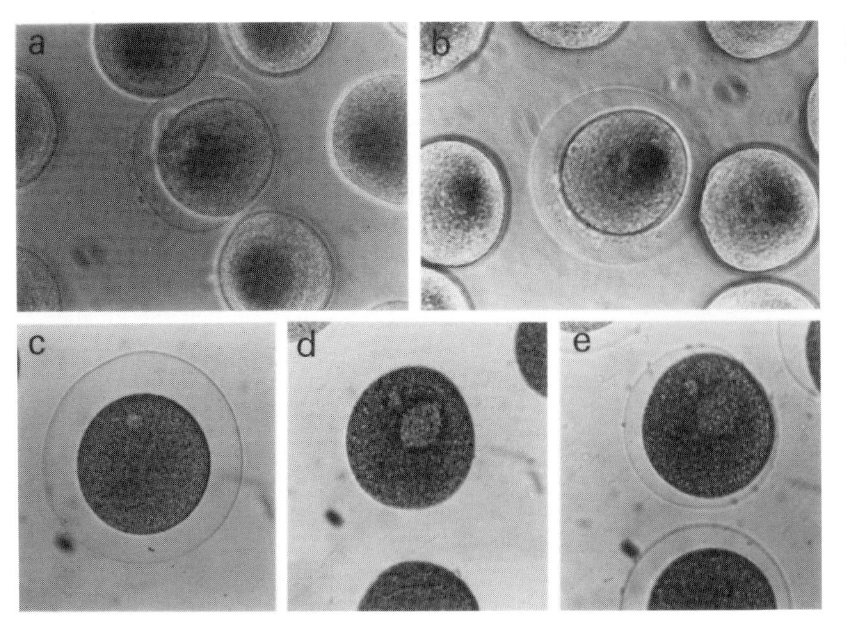

Figure 5.2 The first direct evidence for a soluble sperm factor was obtained by injecting extracts from sea urchin sperm into sea urchin oocytes and observing the elevation of a fertilization membrane **(a** and **b)**. The control injection of a salt solution did not elevate a fertilization membrane **(d)** (from Dale et al. 1985).

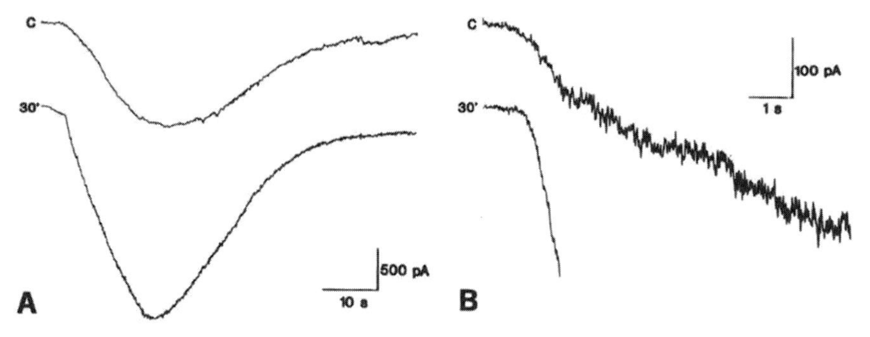

Figure 5.3 Fertilization currents in ascidian oocytes when inseminated with control sperm **(c)** and with sperm pre-treated with lithium (30') to alter the phosphoinositide cycle of the spermatozoon. Note the increase in amplitude and slope of the fertilization current induced by pre-treating spermatozoa showing that an oocyte activation event can depend on the physiological status of the spermatozoon (from Tosti and Dale 1992).

external surface. In the toad, *Xenopus laevis*, fertilization appears to trigger a transmembrane oolemmal protein that triggers a non-SH2 (Src homology 2) mediated activation of a Src-type SFK (Src family kinase) to increase PLCγ activity and IP_3 levels. In the newt *Cynops pyrrhogaster*, the spermatozoa triggers an initial Ca^{2+} spike apparently due to the effect of a protease on the oocyte surface, whereas later calcium waves are generated by an internally acting sperm-borne factor, citrate synthase, which also functions by raising levels of IP_3. The citrate synthase may trigger the calcium waves by increasing levels of PLCγ or alternatively, along with catalysis leading to citrate production, it could also inversely catalyse the cleavage of citrate into acetyl-CoA and oxaloacetate. In birds, which are physiologically polyspermic, the concept of quantity of soluble sperm factor arose, and here the sperm factor was shown to be composed of three factors, PLCζ, citrate synthase and aconitate hydratase. The first is responsible for the initial transient Ca^{2+} increase and the other two for the long-lasting spiral-like Ca^{2+} oscillations which may be due to ryanodine receptor-mediated calcium release. In the large oocytes of *Xenopus*, the amount of IP_3 produced at fertilization increases to a maximum of 555 nm within one minute of insemination. This amount does not vary whether the oocyte is monospermic or induced to be polyspermic with up to 75 spermatozoa entering the oocyte. That is, the binding or delivery of sperm factor from one spermatozoon is as potent as that from 75 spermatozoa. In the polyspermic scenario we would have expected a 75-fold increase in sperm factor and consequently a larger amount of IP_3 to be produced in the oocyte. Sperm–oocyte activation is obviously not so simple. An exciting new hypothesis for oocyte activation could involve the fusion of the membranes and in particular the interaction of the two gametes at the level of lipid rafts and their association with the cortical microfilament system.

There are intrinsic experimental difficulties in all of the techniques used to identify the sperm factor. Injecting sperm extracts can be misleading since spermatozoa contain many factors common to all cells, from Ca^{2+} to IP_3 and NO, that are integral to sperm physiology but not necessarily packaged to serve as an activating factor. In any case, the fact that these common cellular second messengers may trigger events downstream in the complex mechanism of oocyte activation does not mean they are the initial trigger. Gene disruption is a powerful tool to investigate fertilization, however even here results must be treated with caution. For example, the knockout of one factor could be compensated by other similar factors, and some gene products may only interact when together with others. In some cases, gene manipulation may affect other genes located nearby, or there could be an unintentional elimination of microRNAs.

Since changes in plasma membrane conductance, calcium ion release and maturation-promoting factor inactivation are common to all animal oocytes at activation, as they are indeed common components of somatic cells, it would not be surprising to discover that the sperm-activating factors, whether internally or externally acting, are common elements of all cells. There is evidence that the sperm factor is not species specific and that it is more complex than a single specific molecule. In fact, in many animals it has already been demonstrated to be made up of several components that could include both internally and externally acting elements. In any case, as shown by the fact that knockout mice lacking in PLCζ are not sterile, it appears that alternate pathways leading to activation exist in the oocyte (Figure 5.4).

Electrical Events at Activation

The importance of ion fluxes across the plasma membrane in the process of oocyte activation has been recognized for over 70 years. Direct measurements of membrane voltage during oocyte activation using intracellular microelectrodes have been carried out in ctenophores, echinoderms, ascidians, annelids, teleosts, amphibians and mammals. With the advent of patch clamp and whole cell clamp technology, direct measurements of the currents generated at fertilization have been possible while allowing both the simultaneous microinjection of inhibitors and stimulants into the cell. The fertilization current in many invertebrates is a long, bell-shaped inward current that may reach 1,000 pA over a 60-second period that may drastically change the sub-cortical concentration of free cations to an extent much larger than the free Ca^{2+} generated by internal waves (10 μm), and, as we shall see later, free Ca^{2+} itself may sensitize the calcium stores in the oocyte.

The Initial Step Depolarization and the Latent Period

Close observation of the initial phase of the activation potential in sea urchin oocytes in my laboratory in Naples in the 1970s showed it to be composed of discrete step-like events. Each spermatozoon that entered the egg induced a small l–2mV step-like depolarization. A composite shoulder phase indicated polyspermy. In a monospermic situation, of the hundreds of attached spermatozoa, only the fertilizing spermatozoon is capable of reacting with the oocyte, thereby inducing a single step-like depolarization. This report was the first to show how the fertilizing spermatozoon differed from the supernumerary spermatozoa in its capacity to generate a discrete electrical event. This step, which is the earliest detectable event in the oocyte at fertilization, is not seen in parthenogenetically activated oocytes. Voltage clamp studies confirmed this sperm-induced electrical event in the sea urchin, and a comparable event has been identified both in ascidians and in the anuran *Discoglossus pictus*. Figure 5.5 shows that in all these deuterostomes, the step event lasts about 5–10 seconds at room temperature.

Owing to the rapid succession of change in the oocyte during activation, it is difficult to dissect out the cause of the initial step with the subsequent activation events. Germinal vesicle stage oocytes may also be fertilized. However, since they are immature, they do not give rise to the cortical reaction, the fertilization potential, or other autocatalytic events seen in the mature oocyte. When a GV oocyte is fertilized, not all of the spermatozoa are capable of penetrating the cell or of producing a fertilization cone. Over a thousand spermatozoa

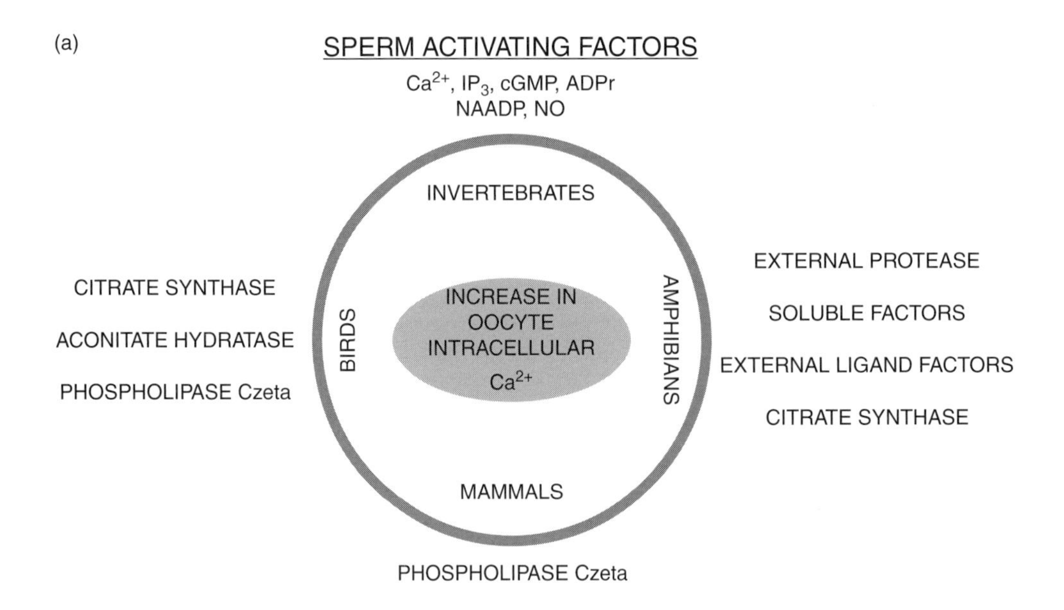

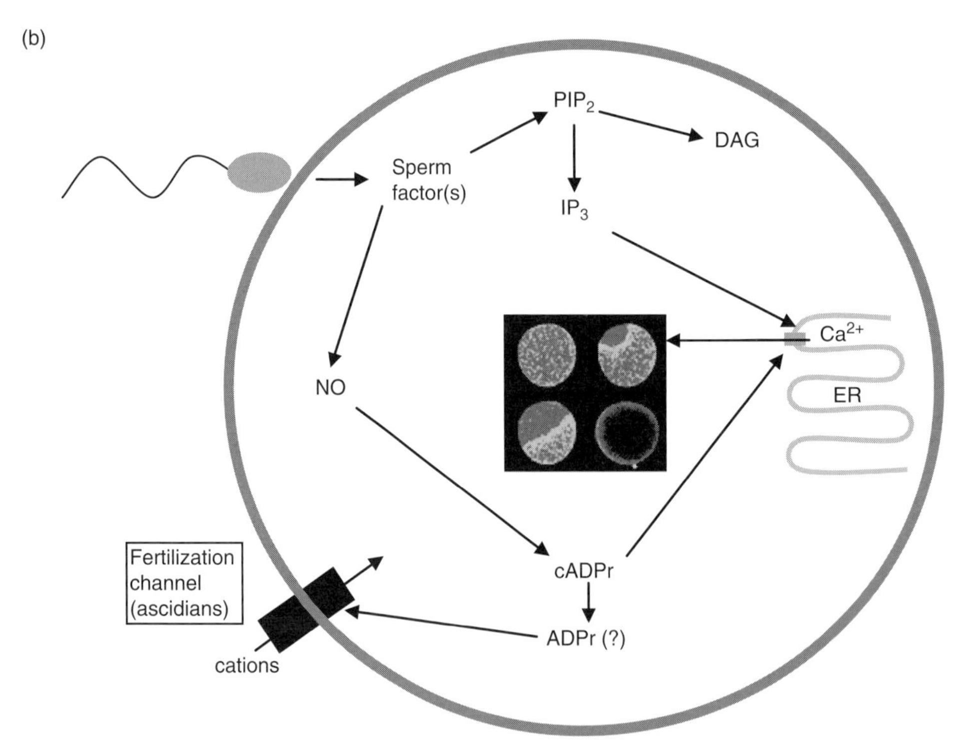

Figure 5.4 **(a)** A summary of proposed sperm-activating factors across the animal kingdom. **(b)** An example of sperm activating factors in the ascidian.

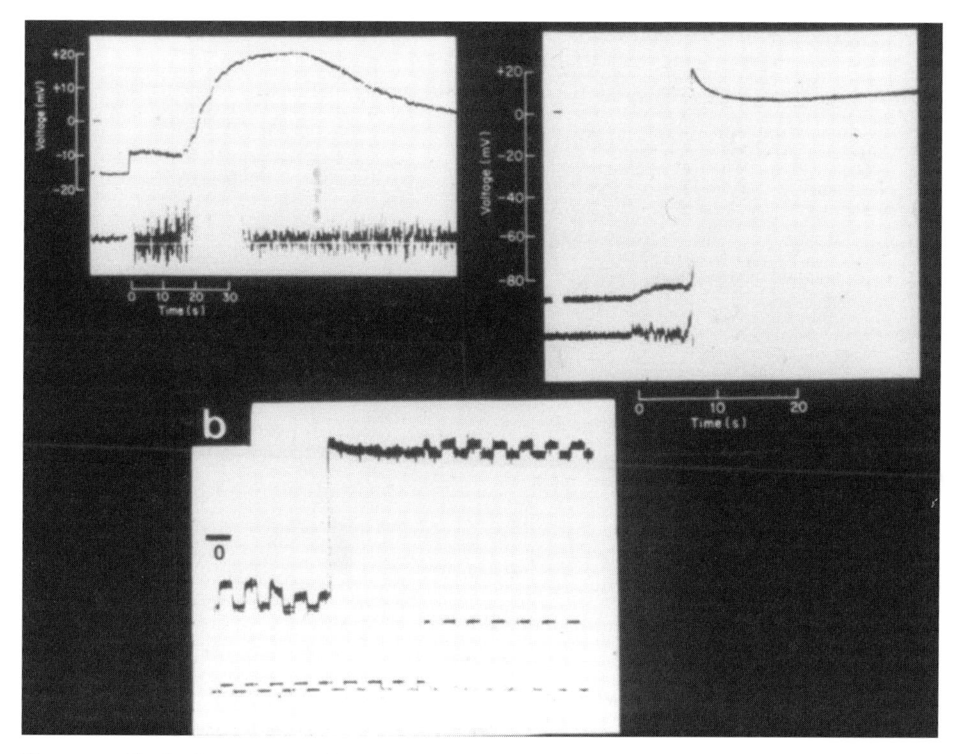

Figure 5.5 The fertilization potential in a sea urchin (top left), an ascidian (top right) and a frog (bottom). Note that in all cases, the larger voltage change is preceded by a smaller step change that indicates sperm–oocyte fusion.

may attach to the GV oocyte, but usually not more than 10 enter, as demonstrated by the formation of fertilization cones (Figure 5.6). If we record electrically from immature oocytes, successful spermatozoa give rise to an electrical depolarization and conductance increase and induce a fertilization cone. Other sperm are not capable of inducing either an electrical event or a cone, whilst a third category induces a step depolarization that after several seconds spontaneously reverses (see also Figure 6.3). These sperm did not enter the oocyte, nor did they induce the formation of a fertilization cone. These experiments raised the possibility that the step depolarization was the direct result of sperm–oocyte fusion, the conductance increase being due to the appearance of sperm channels in the newly formed syncytium. If fusion is inhibited by the ATPase inhibitor quercetin, or by removing Mg^{2+}, spermatozoa are unable to generate electrical changes in the oocyte. Furthermore, the step event may be experimentally reversed by adding a spermicide to inseminated oocytes. Spontaneously or experimentally induced reversible sperm steps have since been seen in a variety of circumstances.

Frank Longo and colleagues voltage-clamped sea urchin oocytes, fertilized them and then serially sectioned them to locate the fertilizing spermatozoon. Since the authors did not detect gamete fusion until five seconds after the step depolarization, they suggested that the step is a pre-fusion event, suggesting that factors in the sperm plasma membrane cause channel openings in the oocyte through a second messenger mechanism. These second messenger systems could be cytoplasmic, such as inositol trisphosphate or ADP ribose, or membrane linked, such as G-proteins. Since G-protein activators closely mimicked events of activation, these authors proposed a G-protein linked mechanism for the generation of these depolarizations. The onset of the step event is coincident with an increase in capacitance. Since, in biological membranes capacitance is proportional to surface area, an increase in capacitance at this moment indicates gamete fusion. To estimate the elementary conductance change underlying the step depolarization in sea urchins, the ratio of the change in voltage noise variance to the change in potential was calculated. Knowing the cell membrane resistance during this change, the single channel conductance

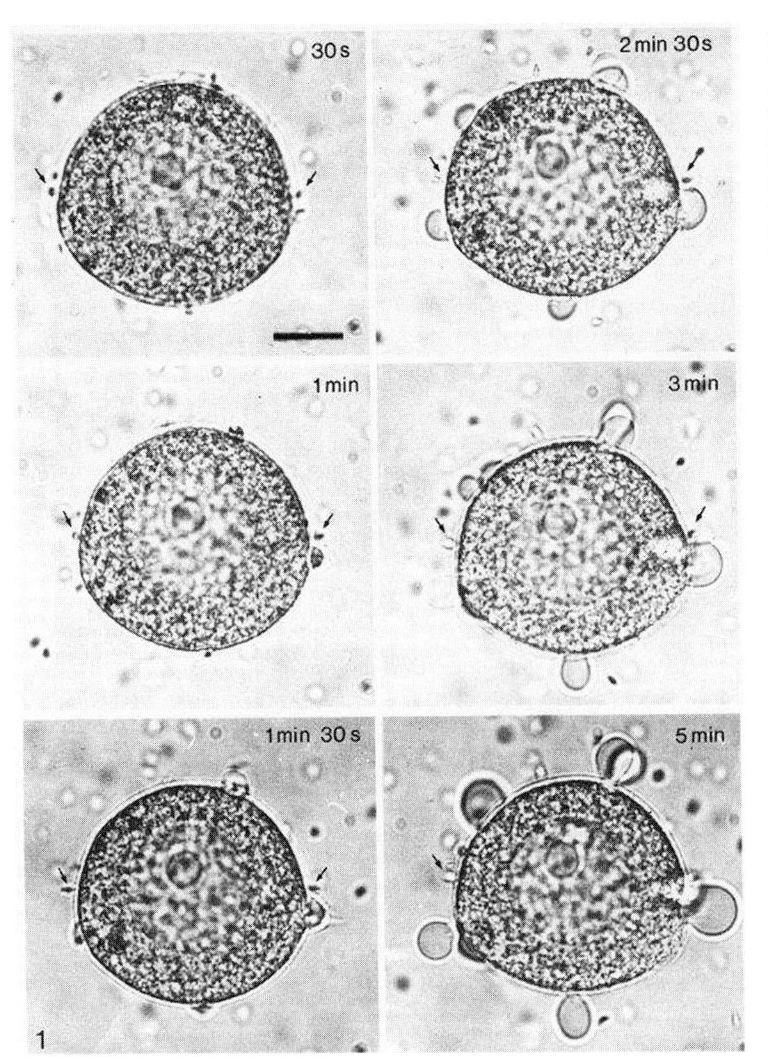

Figure 5.6 A time sequence of sperm–oocyte interaction in the GV oocyte of the sea urchin *Sphaerechinus granularis*. Note that some of the spermatozoa do not induce cytoplasmic protrusions (arrows). Spermatozoa that successfully enter the oocyte generate a step depolarization, stiffen perpendicular to the oocyte surface and then form a cytoplasmic protrusion, which is equivalent to the fertilization cone in the mature oocyte.

was estimated to be about 30–90 pS. It is still not clear if the step event is the result of sperm–oocyte fusion, or the release of a channel gating factor into the oocyte cytoplasm, however since the input resistance of the oocyte decreases from 25 MOhms to 15 MOhms at the step and, assuming the sperm membrane is in parallel with the oocyte membrane, then the sperm would have a conductance of 15 nS, or in other words, it would have to contain 40 channels of 400 pS each to induce the step event.

In most animals, there is a time delay from the moment the spermatozoon attaches to the oocyte surface until cortical exocytosis. The delay between the two corresponds more or less to what has been described in sea urchins as the latent period, during which there is no obvious morphological change in the oocyte surface. The sea urchin oocyte is a useful model to study the latent period for two reasons; first because the latent period is relatively long and is temperature dependent, and second because the cortical reaction may be reversibly interrupted by a mild heat shock, giving rise to 'partially fertilized oocytes'. In these oocytes, 50 per cent of the surface may be activated, while the rest is indistinguishable from a virgin oocyte. Upon re-insemination, spermatozoa are able to interact with the 'virgin surface' of such oocytes, and therefore it appears unlikely that any major change has traversed this area during the latent period. When fertilization occurs in the presence of the microfilament-inhibiting agents cytochalasin B or

D, the latent period is increased by up to 100 per cent. In contrast, there is no change if the gametes are pre-exposed to these agents and subsequently fertilized in natural seawater. Together these experiments suggest that a microfilament dependent stage of sperm–oocyte interaction occurs during the latent period. If the preceding arguments are correct this event is post fusion.

The Fertilization Potential

Mature oocytes are low conductance cells containing a wide variety of both voltage-gated and chemically gated channels. Approximately 5–10 seconds after the step depolarization, there is a much larger depolarization called the 'fertilization potential'. At about the same time, there is a massive release of $Ca^{2+}$from intracellular stores and followed by a cortical reaction leading to cortical granule exocytosis in the sea urchin or a surface contraction in the ascidian.

The potential remains at a positive value for several minutes and then gradually returns to its original value. Recording voltage with an intracellular microelectrode allows a qualitative measurement of the electrical event. To quantify and define the actual molecular mechanism underling these voltage changes, it is necessary to voltage-clamp the membrane with a second intracellular microelectrode or use the patch clamp technique. Using this technique in my laboratory in Naples in the 1980s, a new population of chemically gated channels called 'fertilization channels' were identified in the ascidian oocyte plasma membrane at fertilization.

Fertilization channels in the ascidian *Ciona intestinalis* were found to have a single-channel conductance of up to 400pS. Since the reversal potential was around 0 mV, it was suggested that these channels were not ion specific. To date, these channels are amongst the largest observed in biological membranes. Whole-cell currents in ascidian oocytes also studied during fertilization were shown to peak near -30 mV and approach zero near 0 mV, supporting the single-channel data. Knowing the total conductance change at fertilization, the single channel conductance and the probability of a channel being open, it was estimated that the fertilizing spermatozoon opens between 200 and 2,000 fertilization channels in the oocyte. The fertilization current is a long, bell-shaped inward current of about 1000 pA over a 60-second period. This means that a total charge of 10^{-9} Colombes per second may flow into the oocyte at activation, that if localized and composed mainly of Na^+ and Ca^{2+}, may lead to the entry of 10^{10}ions (Ca^{2+} carries twice the charge of Na^+, but this is ignored in order of magnitude calculations). If this number of ions were to localize to 1 per cent of the oocyte volume in the sub-cortical cytoplasm, the concentration would be 10 mM. Thus, the fertilization current alone may drastically change the sub-cortical concentration of free cations.

Nude ascidian oocytes may be cut into small fragments and each fragment has the capability of developing into an embryo. By using the whole-cell clamp technique on unfertilized and fertilized fragments and inseminating each fragment, it was found that fertilization channel precursors and voltage-gated ion channels are uniformly distributed around the ascidian oocyte surface. Since fertilization currents were similar in whole oocytes or fragments, irrespective of their size and global origin, it was concluded that the fertilizing spermatozoon opens a fixed number of fertilization channels limited to an area around its point of entry. The localized ion current through these channels may regulate movements of the cytoskeleton involved in cytoplasmic segregation. Applying ascidian sperm in a large pipette to the ascidian oocyte surface induces small local fertilization currents that do not cause a contraction, but then if spermatozoa are added to the seawater, a large fertilization current is generated, followed by a surface contraction. This implies that although fertilization channels may be activated all over the oocyte surface, there is a regionalized site at the vegetal pole with a high density of these channels and a pacemaker region where the calcium wave and surface contraction wave are initiated.

In the sea urchin, the fertilizing spermatozoon triggers an inward current of about -500 pA, while the conductance increases from 20 to 40 nS. The I/V curve for this current and the reversal potential of about +10 mV suggest it is also non-specific for ions. As mentioned there are also several types of voltage-gated ion specific channels in oocytes, and these may also be activated during fertilization; however, the reversal potential for Ca^{2+} and Na^+ channels does not coincide with the reversal potential for the fertilization current.

In amphibians, a Cl^- specific channel is responsible for the fertilization potential. For example, the membrane depolarization at fertilization in *Xenopus laevis* is influenced by the external chloride concentration and this is a consequence of Cl^- ion efflux. Similarly, a positive shift in the fertilization potential was shown in *Rana pipiens,* which was associated with an increase of either K^+ or Cl^- and a decrease in Na^+ conductance. Membrane potential changes in *Rana cameranoi* oocytes are also based on K^+ as well as on Cl^- conductance. Cl^- ions are apparently responsible for the first depolarization phase evoked by sperm, whereas K^+ contributes to the repolarizing phase. The role for Ca^{2+} channels in the amphibian fertilization potential is less clear.

In the anuran *Discoglossus pictus* the sperm entry site is a predetermined specialized area called the animal dimple. *Discoglossus* oocytes, as other frog oocytes, may be activated by pricking with a steel needle. (It should be pointed out that oocytes from other animal groups are not activated upon pricking.) At fertilization, the spermatozoon initiates a large regenerative depolarization that is Cl^- dependent; pricking elicits a comparable response. When oocytes were pricked outside the dimple area, a wave of contraction spread from the puncture site to the antipode. Since the activation potential was not generated until the wave reached the dimple, it appears that the channels underlying the depolarization are found in the dimple, and that they are gated by a second messenger liberated in the oocyte cytoplasm that spreads around the oocyte with the contraction wave (Figure 5.7).

The fertilization potential in mammals is similar in time course to those in invertebrates and amphibians, however inverted in polarity. The first measurements of membrane potential at fertilization in the hamster and mouse showed a series of hyperpolarizations, while subsequent studies showed the underlying fertilization current to be an outward Ca^{2+} activated K^+ current in the human oocyte (Figure 5.8). The pattern in rabbit oocytes is different, in that a preliminary depolarization is followed by a repeated diphasic hyperpolarization/depolarization pattern of membrane activity.

Gating of the Fertilization Channels

Fertilization channels are specialized ligand-gated ion channels. To date, two ligands have been identified, Ca^{2+} and ADP-ribose. In ascidians, the fertilization channels are not Ca^{2+} gated, but by ADP-ribose. In fact, raising the level of intracellular Ca^{2+}in ascidian oocytes, by perfusion or by loading the oocyte cortex (>50 μm) with Ca^{2+}through voltage gated channels, did not activate fertilization channels. Alternatively, oocytes exposed to low-Ca^{2+} seawater, perfused with the Ca^{2+} chelator K-EGTA or Ca^{2+} blocking agents to prevent the release of Ca^{2+} from intracellular organelles, and subsequently inseminated, generated fertilization currents. Oocytes exposed to the Ca^{2+} ionophore A23187 were found to contract without generating a fertilization current, while microinjection of IP_3 or soluble fractions of homogenized spermatozoa induced both a contraction and a fertilization current. Although elevated Ca^{2+} does

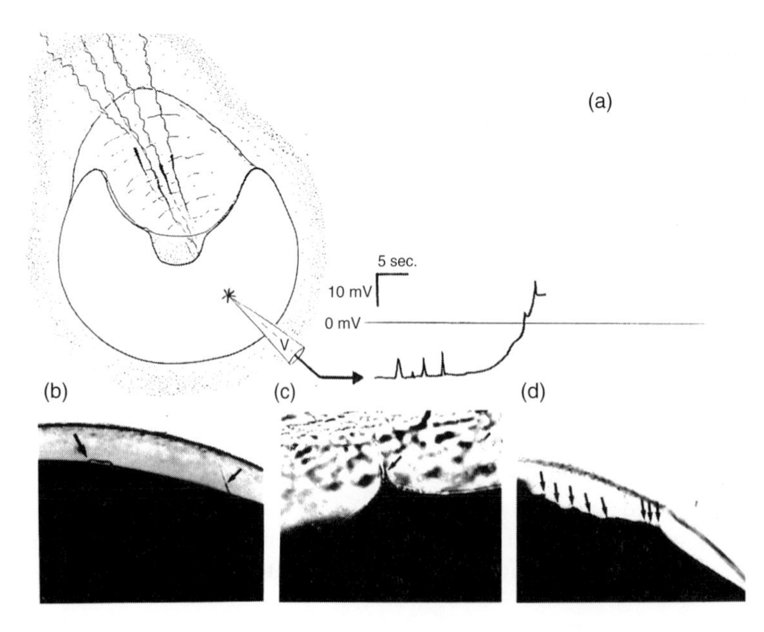

Figure 5.7 In the painted frog *Discoglossus pictus,* the fertilization potential is generated by the activation of Cl^- channels localized in the dimple (courtesy of Professor Chiara Campanella).

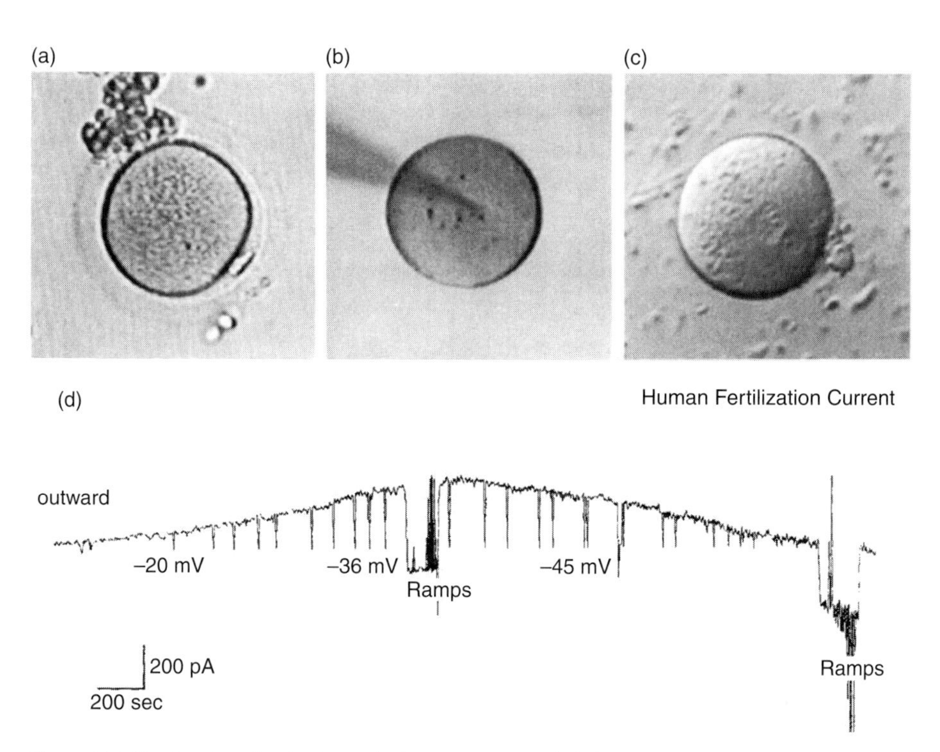

Figure 5.8 The first sign of activation in the human oocyte is a bell-shaped outward current generated by the activation of Ca^{2+} activated K^+ channels. The frames at the top show the oocyte **(a)** before recording, **(b)** during recording and **(c)** 18 hours after the registration, showing the pronuclei (from Gianaroli et al. 1994).

not gate fertilization channels in ascidians, it appears to be involved in the mechanism of cortical contraction. Measurements with ion-selective electrodes show that the intracellular pH of ascidian oocytes ranges from 7.2 to 7.4, and this does not vary during activation, making pH an unlikely trigger of early activation events.

Nitric oxide has been shown to increase in ascidian oocytes at fertilization. Generating cytosolic nitric oxide in ascidian oocytes with the donor sodium nitroprusside triggers a fertilization-like current and the release of intracellular calcium through a ruthenium-red sensitive mechanism. At the same time, microinjection of soluble extracts of ascidian sperm causes calcium release in ascidian oocytes but not gating of the fertilization channel. With whole cell and single channel recording, it has been shown that the fertilization channel is directly gated by ADP-ribose (Figure 5.9). The channel was shown to be permeable to Ca^{2+} and Na^+, with a reversal potential of 0 to +20 mV, and a unitary conductance of 140 pS. BAPTA or antagonists of intracellular calcium release did not inhibit the ADP- ribose current, showing it is activated in a calcium-independent manner. In situ, the fertilization current is blocked by nicotinamide. In conclusion, ascidian sperm trigger the hydrolysis of nicotinamide nucleotides in the oocyte to ADP-ribose, which directly gate the fertilization channels, and also assists in the release of intracellular Ca^{2+}. Since sperm extracts do not trigger all activation events, it seems probable that there are multiple pathways at activation that are gated by more than one factor from the spermatozoon.

In the sea urchin, the fertilization current may be induced by microinjection of IP_3 and inhibited by BAPTA and therefore probably Ca^{2+} gated, while in amphibian oocytes, there is evidence that intracellular Ca^{2+} contributes to gating the Cl^- channels. The fertilization current in *Xenopus* oocytes is generated by Ca^{2+} activated Cl^- channels. In the human, the fertilization channels are Ca^{2+} activated K^+ channels. Loading human oocytes with EGTA prevents the fertilization current, while increasing Ca^{2+} by injecting IP_3 or adding the calcium ionophore induces the fertilization current (Figure 5.10). The oscillations in membrane potential coincide with the oscillations in intracellular calcium that continue up to pronuclear formation. These

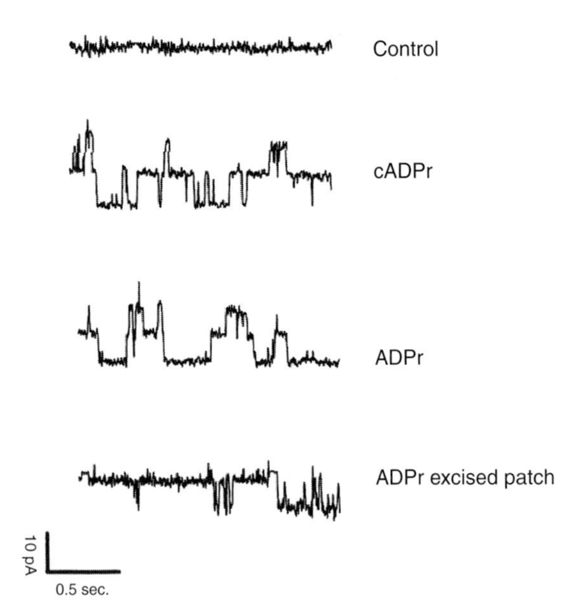

Figure 5.9 Fertilization channels in the ascidian oocyte are gated by the second messenger ADPr. The recordings are single channel recordings using the patch clamp technique showing the fertilization channel has a conductance of 140 pS (Wilding et al. 1998).

oscillations decrease in frequency and amplitude during their progression. The oscillations are large in hamsters and smaller in mice and rabbits. After oocyte activation, plasma membrane Ca^{2+} channels play a role in replenishing stores for the continuation of Ca^{2+} oscillations; in fact in hamsters, external Ca^{2+} is required for oocyte activation. In the bovine, a clear relationship between the electrical properties of the oocyte plasma membrane and intracellular calcium modifications has also been shown following fertilization, as well as following chemical oocyte activation or after exposure to specific Ca^{2+} mobilizers.

Calcium Events at Activation

Daniel Mazia in 1937, was the first to describe an increase in free Ca^{2+} upon fertilization of sea urchin oocytes, however this increase was not seen directly until 1977 when Ridgway, using a calcium-sensitive dye, demonstrated it in the large oocytes of the fish medaka. This increase in intracellular calcium is due to the release of calcium ions from cytoplasmic stores into the cytoplasm that occurs simultaneously or shortly after the plasma membrane currents are triggered. Ca^{2+} release occurs either in a single transient peak, as in many jellyfish, echinoderms, molluscs, nematodes, fish and amphibians; or in a series of oscillations as seen in ascidians, mammals and some annelids and arthropods (Figure 5.11). In all cases the amplitude and time course are crucial for successive activation events and embryo development. In many animals, including the starfish and sea urchin, the calcium wave is preceded by a global increase in sub-cortical calcium, called the cortical flash, that may be due to the entry of calcium through plasma membrane channels. In the golden hamster and ascidians, periodic hyperpolarizations of the plasma membrane are associated with the intracellular Ca^{2+} transients. The intracellular store for Ca^{2+} in oocytes appears to be the endoplasmic reticulum, where both ryanodine and inositol 1,4,5 triphosphate (IP_3) receptors are located. The calcium increases occur in waves, starting at the point of sperm entry and traversing the oocyte to the antipode (Figure 5.12). The velocity of these waves has been studied for a variety of animals and generally fall into two categories: fast or slow (Table 5.1). For example, in jellyfish and the anuran *Xenopus laevis*, where the sperm enters the animal pole of the oocyte, the calcium wave starts at the animal pole and traverses the oocyte to the antipode, while in nemertean oocytes, the wave initiates at the vegetal pole, the site of sperm entry. In teleosts, where the spermatozoa are forced to enter at the animal pole through the micropyle, the calcium wave starts at the animal pole. Since in mammals and echinoderms the calcium wave is also initiated at the point of sperm entry, more work is necessary to determine whether in these groups there are also preferential sperm entry sites. Calcium oscillations may be bypassed in some species of mammals by pharmacological and genetic treatments that lead directly to a reduction in MPF, and these oocytes may develop at least to the blastocyst stage. This does not mean the initial events are not necessary, merely that the process has been triggered downstream. What science is trying to fathom is how does the spermatozoon in nature trigger the oocyte into activation.

Activation Events in Deuterostomes

Based on several characteristics of post-fertilization development, the metazoa may be conveniently divided into two categories, the deuterostomia and the protostomia. Although the patterns of development within either group are far from homogeneous,

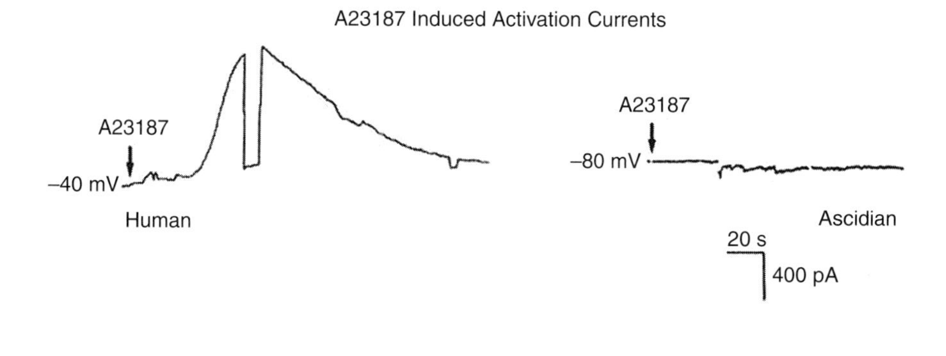

EGTA pre-loaded Oocyte + A23187 or Spermatozoa

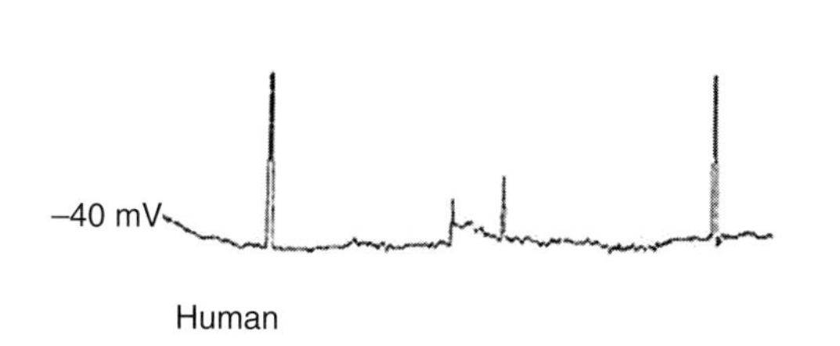

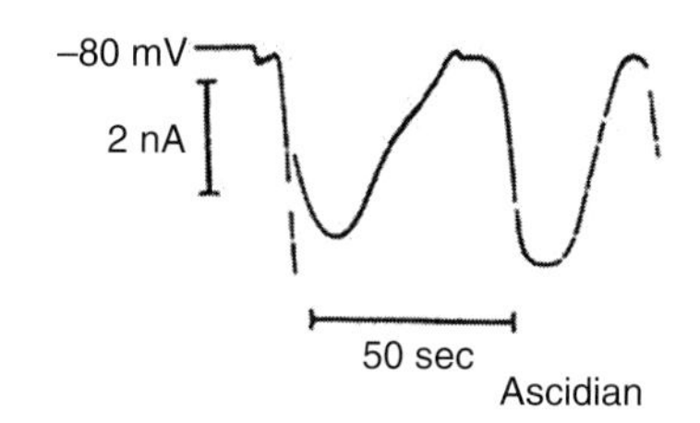

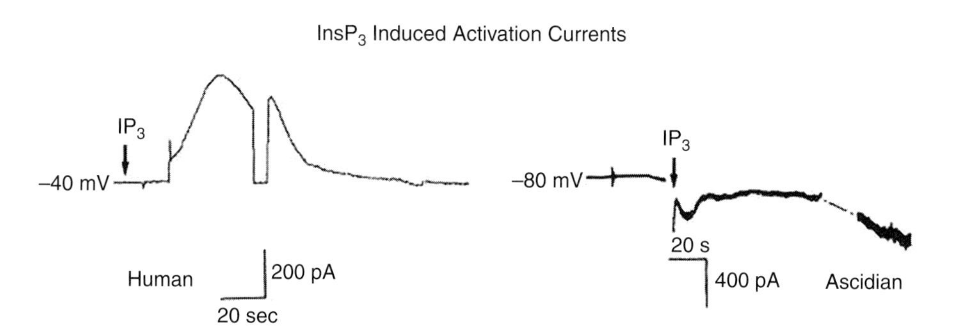

Figure 5.10 Whereas fertilization channels in the human are gated by Ca^{2+}, the equivalent channels in the ascidian are not. The recordings on the left are whole-clamp recordings from human oocytes; those on the right from ascidian oocytes. From the top, following exposure to the calcium ionophore A23187, the human activation current is generated while the ascidian current is not. Centre traces, pre-loading oocytes with the calcium chelator EGTA blocks the sperm-induced current in the human, but not in the ascidian. The bottom traces show that microinjecting the second messenger IP_3 in both oocytes generates the respective currents.

coherent evolutionary trends may be distinguished. The deuterostomes are characterized by radial cleavage, where the axes of early cleavage spindles are either parallel or at right angles to the polar axis and the early blastomeres are indeterminate, i.e. with relatively unfixed fates. In deuterostomes, the mouth arises at a location distant to the blastopore and the mesoderm arises by enterocoelic pouching, where the wall of the archenteron evaginates to form pouches, the cavity of the evagination becoming the coelom and the wall forming the mesoderm. Ascidians, echinoderms and chordates are deuterostomes.

Activation in the Sea Urchin Oocyte

In sea urchin oocytes, which are arrested at G1 of first mitosis at spawning, the first visible indication of activation is a cortical flash of elevated calcium that appears simultaneously around the surface and lasts about eight seconds (Figure 5.13). This first calcium increase is thought to be due to influx through voltage-gated Ca^{2+} channels, and is in fact inhibited by nifedipine. After a latent period of about 15 seconds, which, incidentally, corresponds to the delay period between the two plasma membrane currents, a single calcium wave travelling at 10 μm/sec starts at the site of sperm

Table 5.1 The Speed of Calcium Waves in Oocytes in species with repetitive oscillations, such as mammals and ascidians, the initial wave is slower than consecutive waves.

	Type	Speed
Sea Urchin	Single Transient	10 μm/sec
Starfish	Single Transient	5 μm/sec
Fish	Single Transient	10–15 μm/sec
Frog	Single Transient	5–20 μm/sec
Newt	Single Transient	5 μm/sec
Mammals	Initial Wave	20 μm/sec
Nemetea	Oscillations	10–15 μm/sec
Jellyfish	Single Transient	4–6 μm/sec
Ascidian	Initial Wave Oscillations	7 μm/sec 15 μm/sec

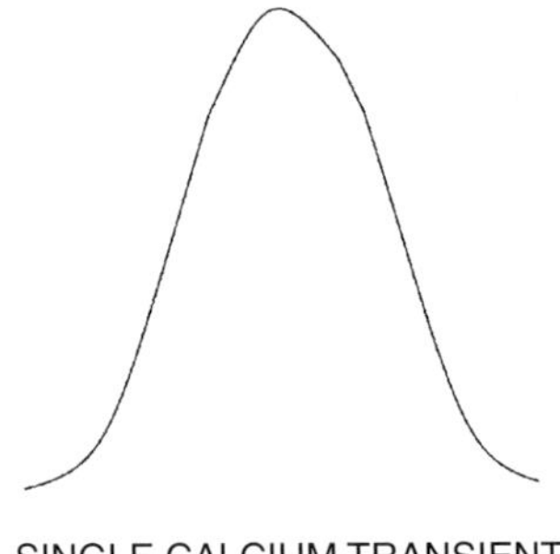

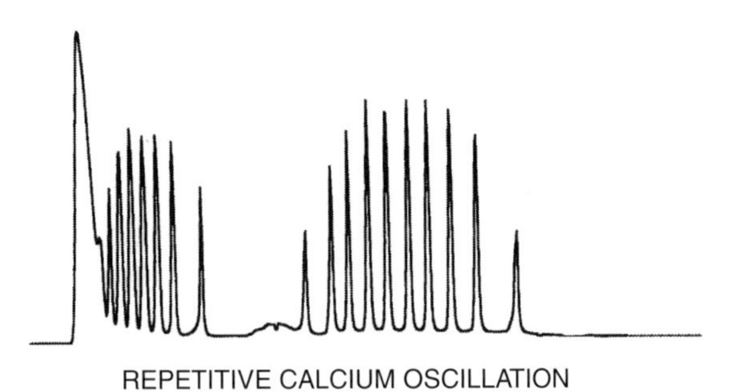

Figure 5.11 Increases in intracellular Ca^{2+} in oocytes induced by spermatozoa may take the form of a single transient, as in echinoderms, or repetitive oscillations, as in ascidians.

entry and traverses the oocyte to the antipode. The cortical flash is dependent on external Ca^{2+}, whereas the calcium wave is due to mobilization of calcium from intracellular stores primarily in the endoplasmic reticulum triggered by IP_3 and cADPr. NAADP also mobilizes calcium in sea urchin oocytes via two-pore channels (TPCs), a new class of Ca^{2+} permeable channel, located on cortically located acidic organelles. This pathway may serve to prime the IP_3 and cADPr sensitive stores in the endoplasmic reticulum. NAADP triggers Ca^{2+} release from the stores that are insensitive to IP_3 and cADPr. Nitric oxide (NO) is also involved in calcium release in sea urchins and is possibly required for the maintenance of elevated Ca^{2+}. One model suggests that SLKs (Src family kinase) are activated at fertilization to trigger PLCγ via an SH2 (Src homology 2) mechanism which hydrolyses PIP_2, generating IP_3 and the Ca^{2+} wave. G-proteins may also be involved in the genesis of the calcium wave by activating the SLKs or by upregulating the cADPr sensitive calcium release. Although the pathway and the interaction of the calcium release intracellular messengers have not been fully elaborated, the effect of the rise in intracellular calcium on MAPK inactivation, CDK2 (cyclin-dependent kinase 2) activation, DNA synthesis and cell cycle progression is clear.

Activation in the Starfish Oocyte

Starfish oocytes are arrested in prophase 1 of meiosis. As for sea urchins, the first calcium response is a cortical flash followed by a single Ca^{2+} wave that travels at a speed of 5 μm/sec (Figure 5.13a). This is followed by erratic multiple calcium transients in the latter part of the cell cycle. The sperm-induced Ca^{2+} signal is therefore biphasic and mirrors the electrical events at the plasma membrane. The cortical flash in fertilized starfish eggs is much smaller and short-lived than that of the sea urchin, lasting about 2–5 seconds. Uncaging of NAADP in immature or mature oocytes induces a Ca^{2+} increase which is uniformly distributed at the periphery of the cells. This cortical flash, which then spreads centripetally, is dependent on the presence of external Ca^{2+}, and appears to be due to influx through the activation of Ca^{2+} channels.

These differences in the cortical flash of these two echinoderms may be due to the different meiotic stages at the time of fertilization or the different structural organization of the cortices.

The main mechanism of calcium release in starfish oocytes is through the hydrolysis of PIP_2 generating IP_3 that opens intracellular calcium channels in the endoplasmic reticulum. In the starfish it appears that cADPr does not sustain the Ca^{2+} wave at fertilization; however, a triggering role for NAADP in the sperm-induced Ca^{2+} response may exist.

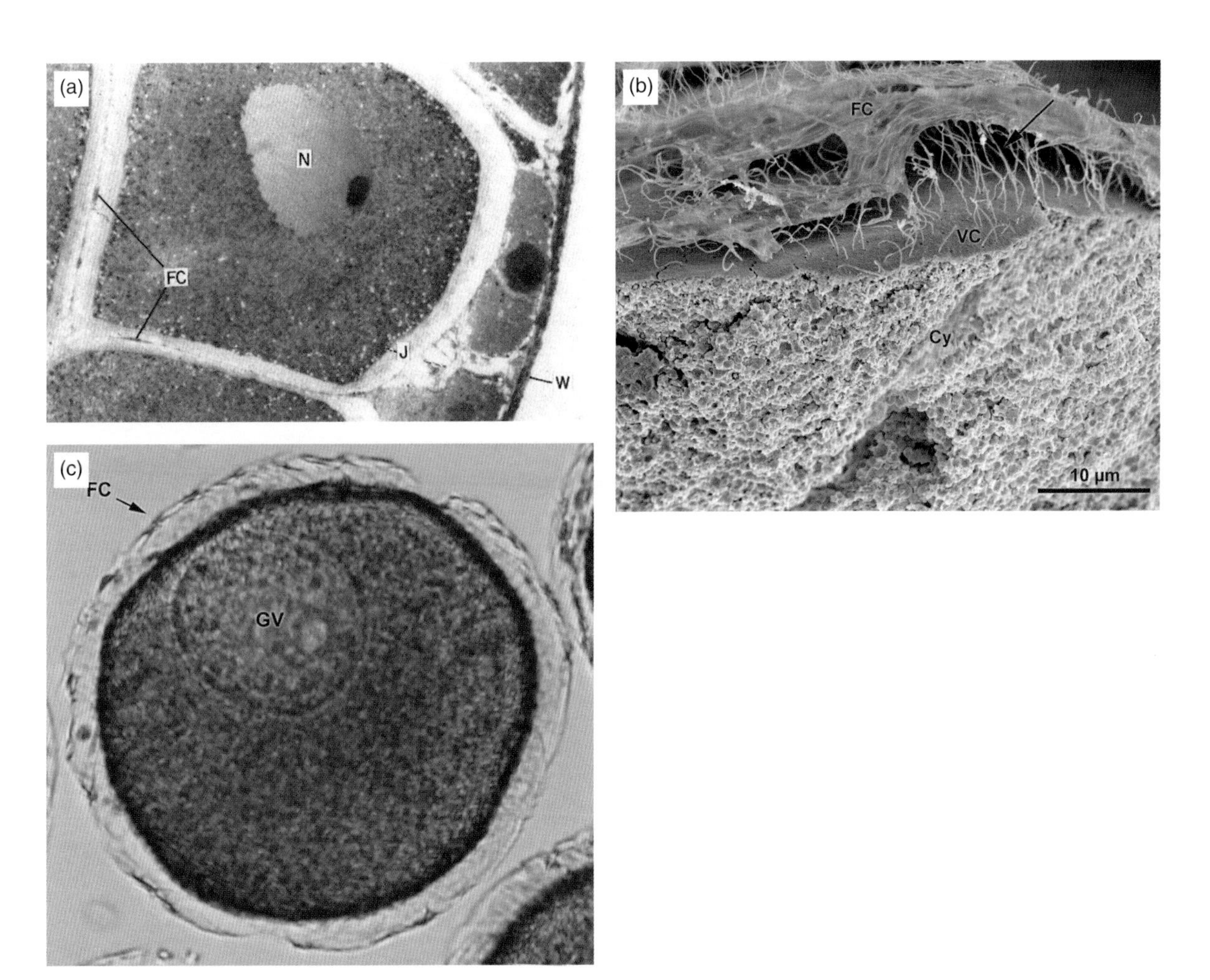

Figure 2.2 **(a)** An oocyte and its follicle cells in situ in the ovary of the starfish *Astropecten aranciacus*. N – Nucleus; FC – follicle cells; J – jelly layer; W – ovary wall. **(b)** A scanning electron micrograph of a fractured starfish oocyte to show the follicle cells (FC) and its projections to the oocyte plasma membrane. VC – vitelline coat. CY – cytoplasm. **(c)** A transmitted light photograph of a starfish oocyte showing the surrounding follicle cells (FC) and the oocytes large germinal vesicle (GV). By courtesy of Dr Luigia Santella, Stazione Zoologica

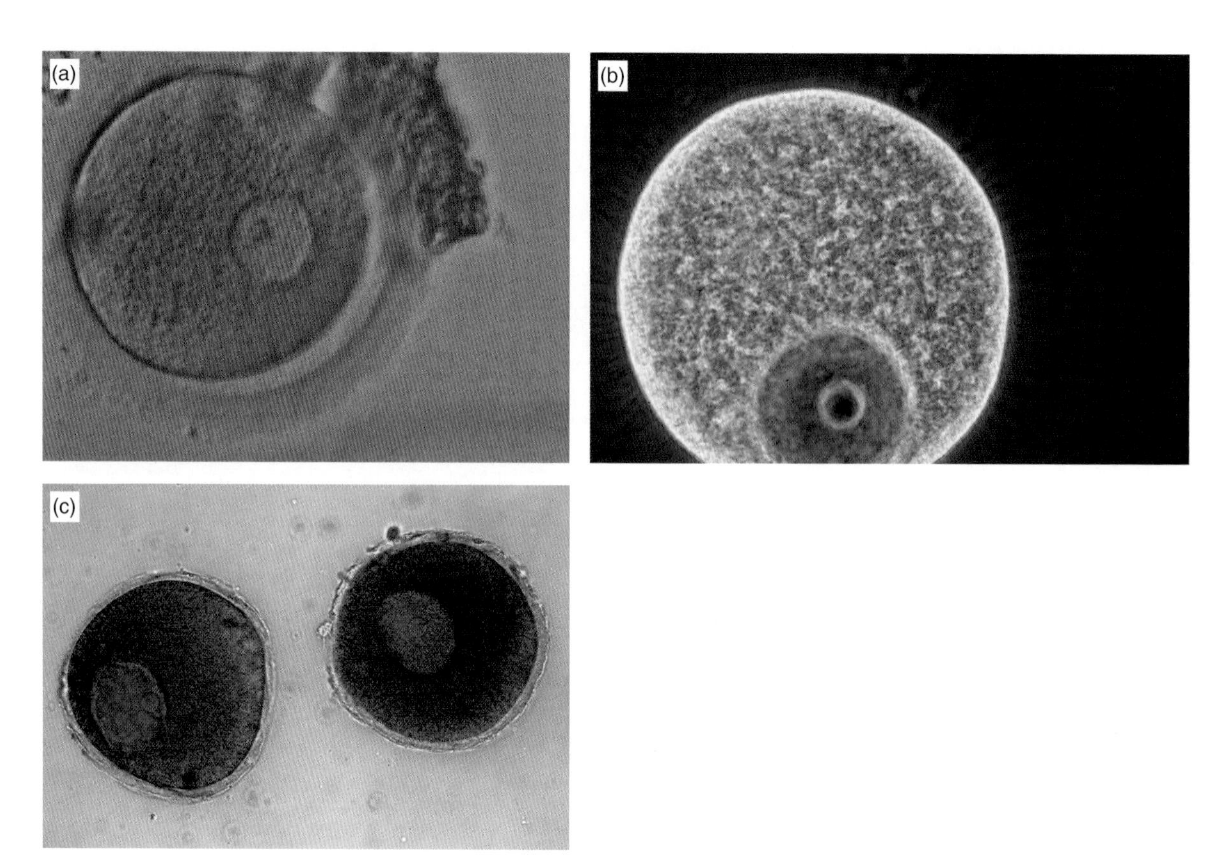

Figure 2.7 The eccentrically located germinal vesicle in Prophase 1 oocytes of the **(a)** human, **(b)** sea urchin and **(c)** starfish. The latter is courtesy of Dr Keiichiro Kyozuka.

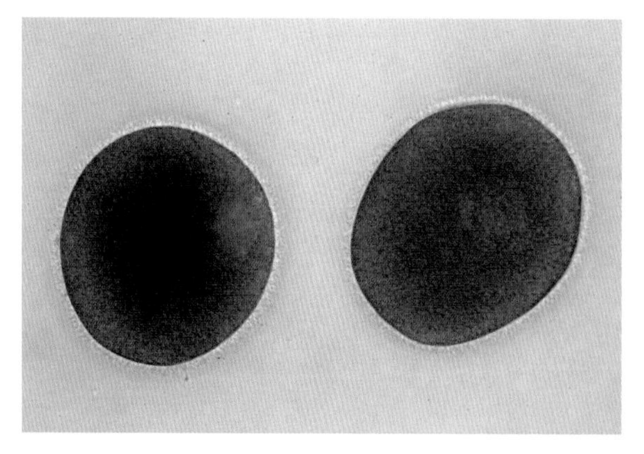

Figure 3.2 Oocytes of the starfish *Asterina pectinifera* 50 minutes after application of the maturing hormone 1-methyladenine (bottom). The oocytes are 160 μm in diameter. The jelly layer is much thinner and more compact than that of the sea urchins (courtesy of Dr Keiichiro Kyozuka).

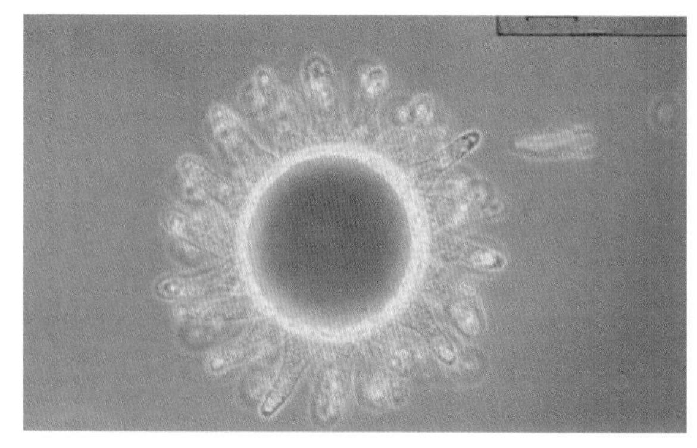

Figure 3.4 An oocyte of the ascidian *Ciona intestinalis* showing the follicle cells, which help in flotation, and the thick extracellular coat called the chorion.

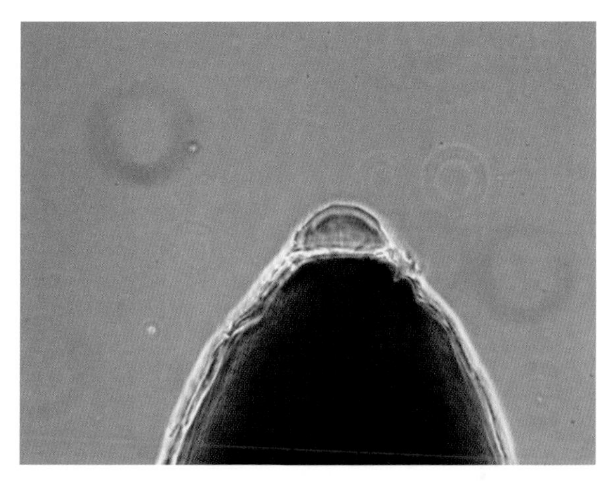

Figure 3.5 The micropyle and jelly cap in the oocyte of the housefly *Mus domestica* (courtesy of Professor Ryuzo Yanagamachi).

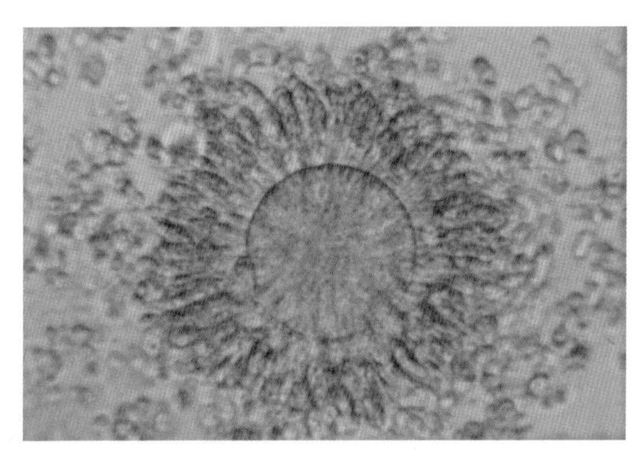

Figure 3.8 The metaphase 2 human oocyte, showing the polar body to the left, the zona pellucida, and the extensive layer of cumulus oophorus, which is made up of cells and hyaluronic acid.

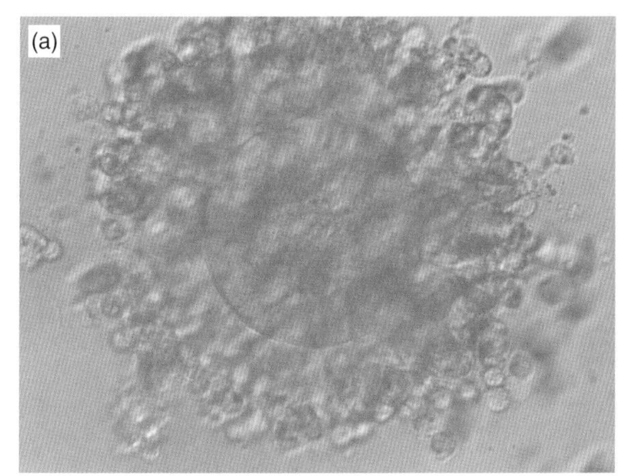

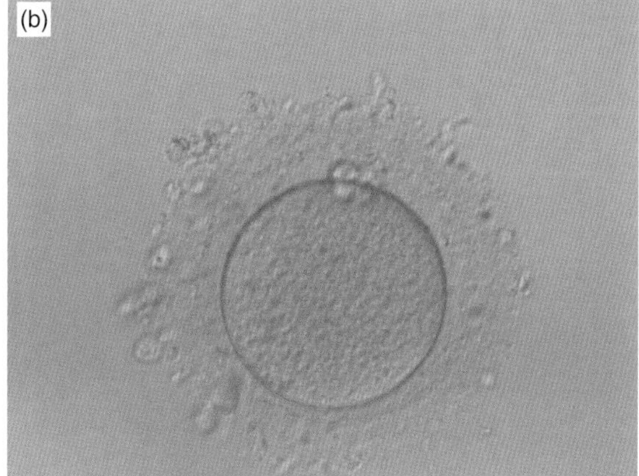

Figure 4.4 **(a)** A human oocyte shortly after aspiration from a follicle showing the extensive and thick cumulus composed of cells and a hyaluronic acid matrix. **(b)** The same oocyte after the cumulus has been removed showing the polar body.

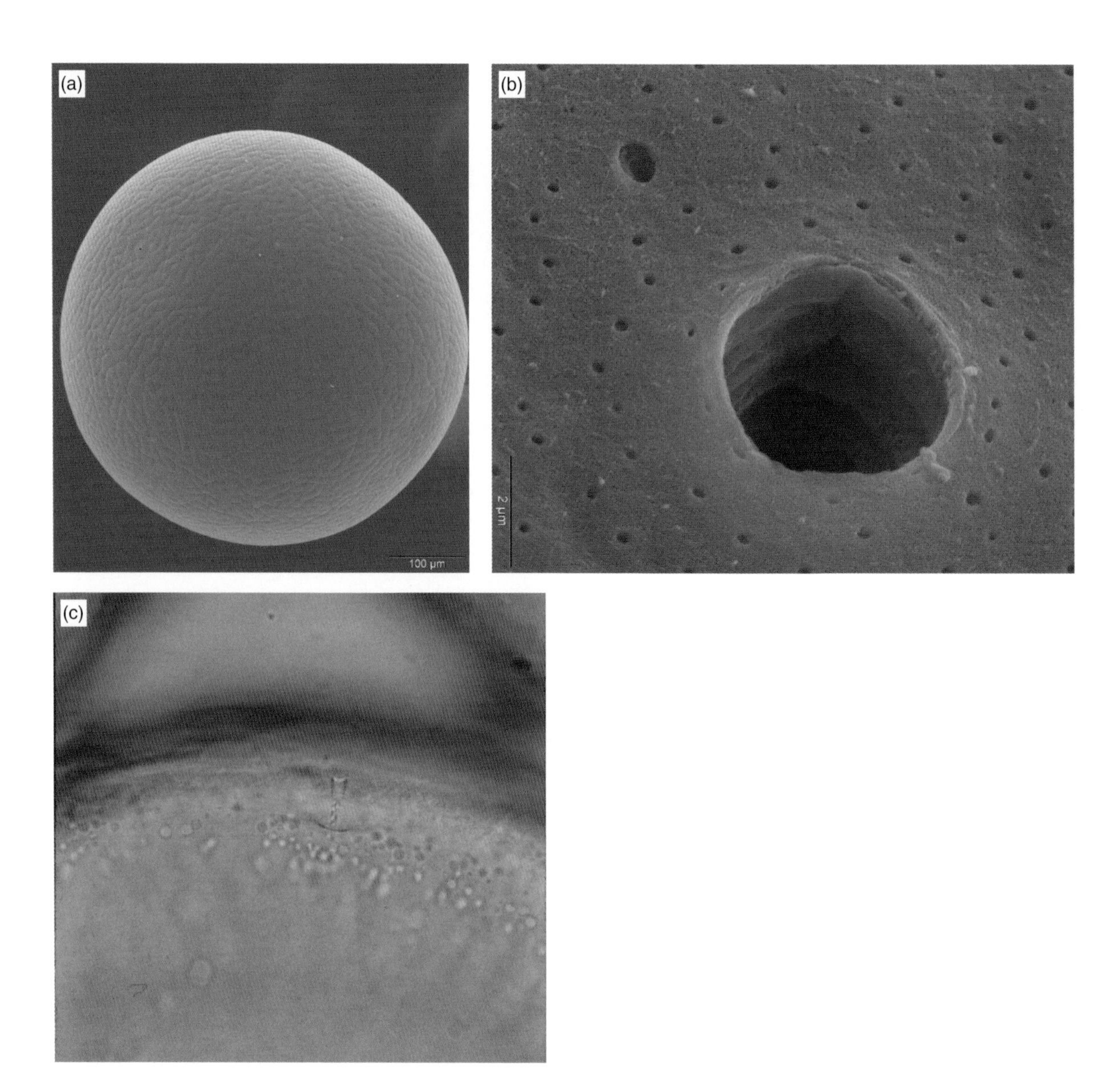

Figure 4.12 **(a)** The micropyle in the oocyte of the black flounder, **(b)** at higher magnification and **(c)** at the light microscope with several spermatozoa in the canal. The canal is 6 μm wide (courtesy of Professor Ryuzo Yanagimachi).

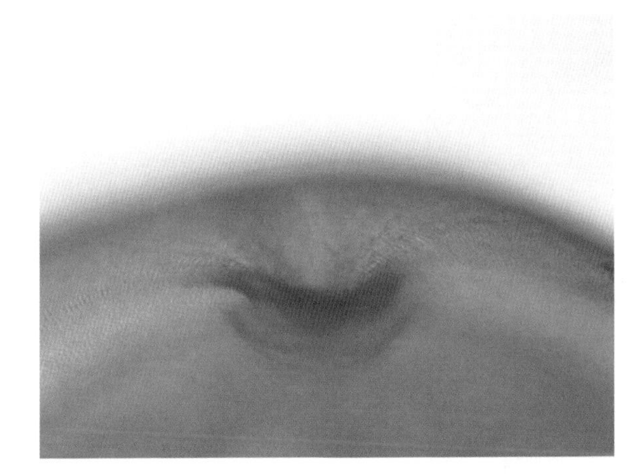

Figure 4.13 The grooved micropyle in the zebrafish (courtesy of Professor Ryuzo Yanagimachi).

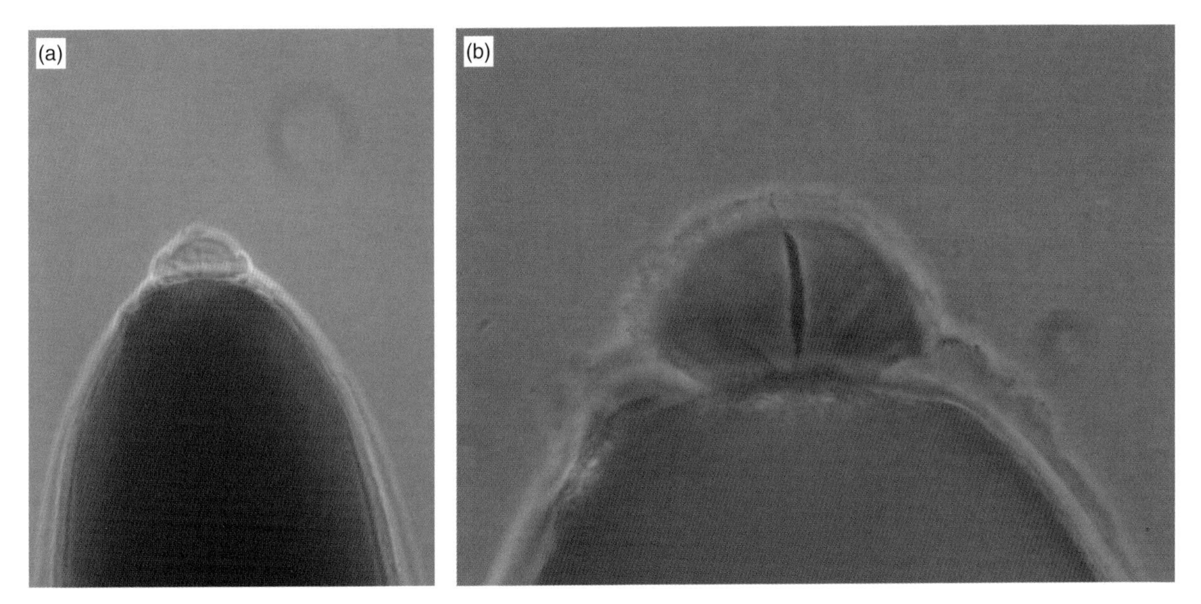

Figure 4.14 **(a)** The jelly cap and the micropyle in the oocyte of the housefly *Mus Domestica* and **(b)** at higher magnification showing a spermatozoon in the micropyle. The jelly cap is some 46 μm wide (courtesy of Professor Ryuzo Yanagimachi).

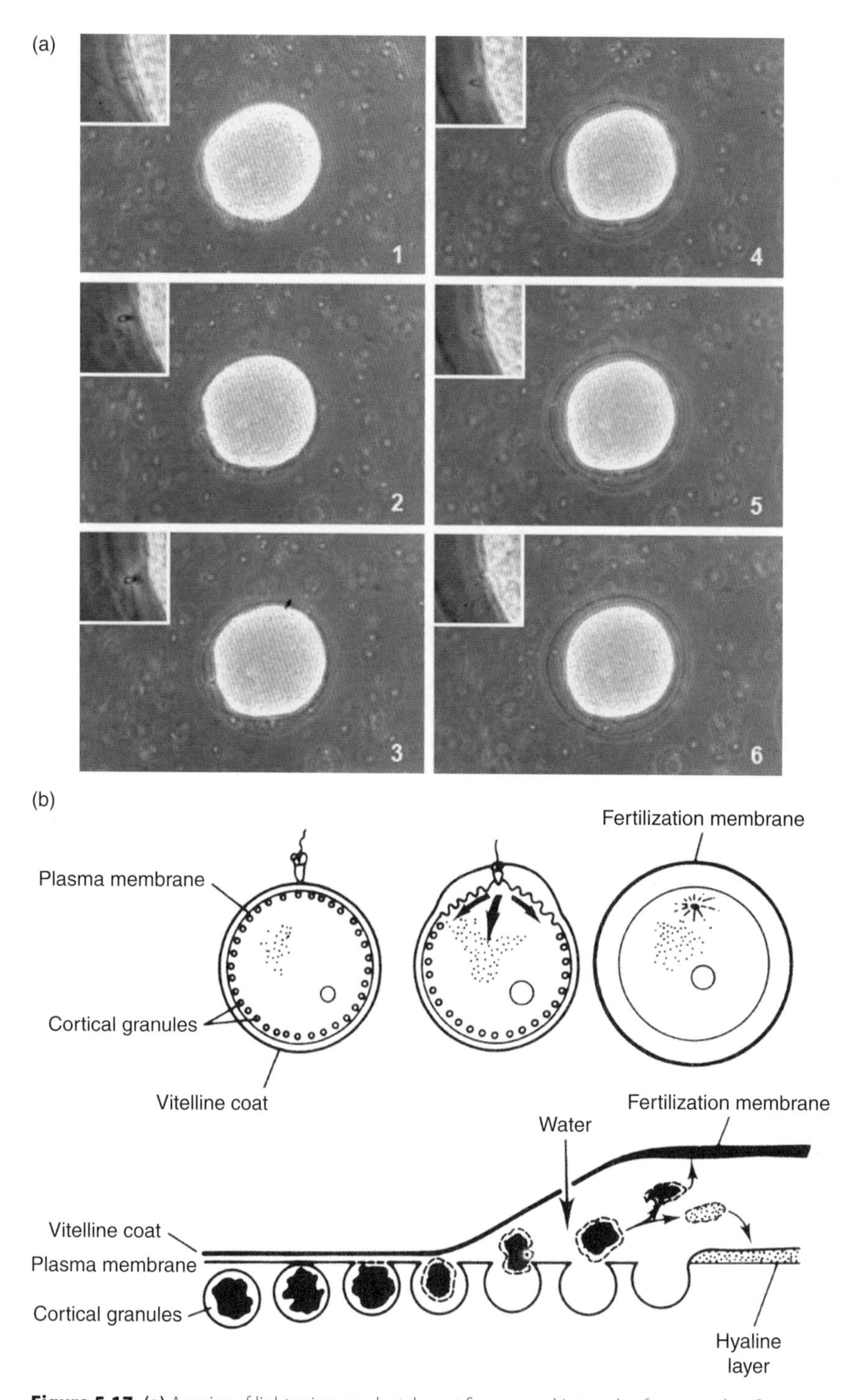

Figure 5.17 **(a)** A series of light micrographs taken at five-second intervals of a sea urchin *Paracentrotus lividus* oocyte showing the entry of the successful spermatozoon (to the left) and at higher magnification in the insets. Note the flattening or concavity at the sperm-attachment site and the elevation of the fertilization membrane in Frame 2 that spreads around the oocyte in a wave, taking about 20 seconds to form a global protective covering. **(b)** A diagram showing cortical exocytosis and resulting elevation of the fertilization membrane in the sea urchin oocyte (from Dale 1983).

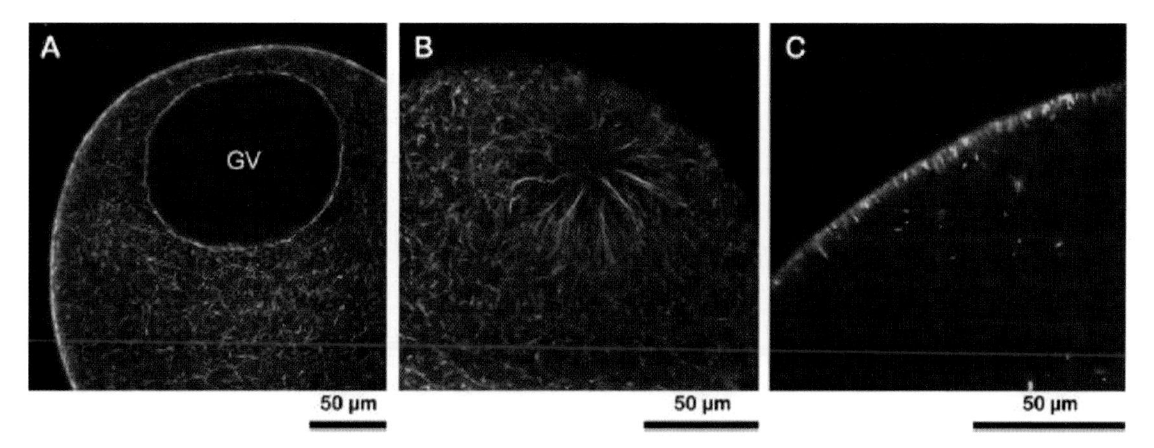

Figure 5.21 Changes in the actin cytoskeleton in the oocyte of the starfish *Astropecten aurantiacus* during maturation. **(a)** The GV stage showing a general dense distribution of F-actin in the cortex, **(b)** the radial arrangement of F-actin fibres over the surface of the germinal vescicle and **(c)** the perpendicular spikes of actin fibres reformed in the cortex of the mature oocyte (from Santella et al. 2015).

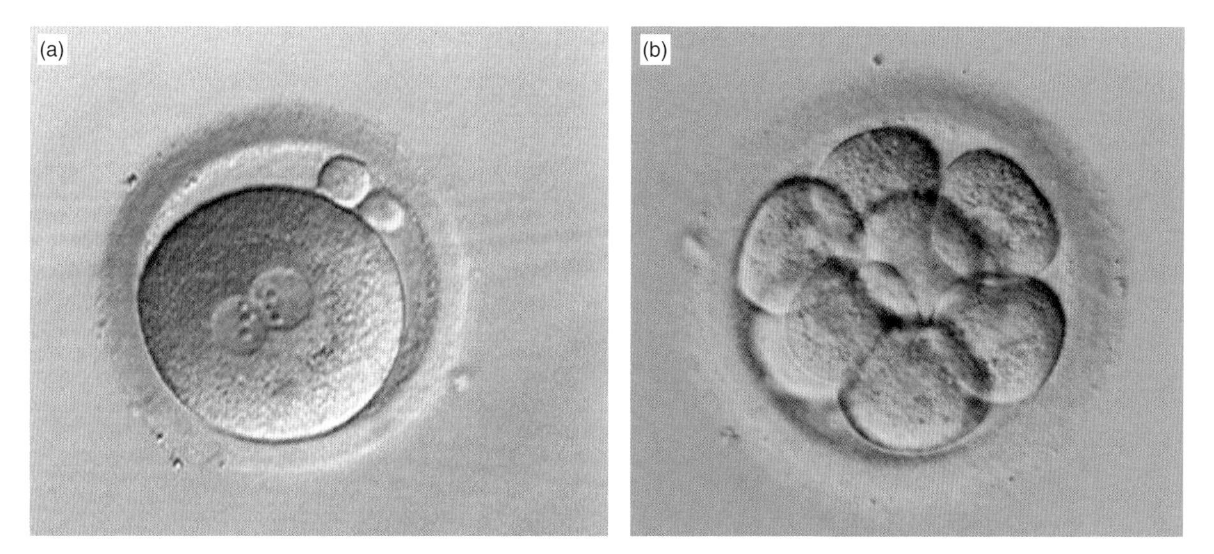

Figure 7.11 The zygote divides by mitotic divisions, called cleavage, into smaller cells the blastomeres. Cleavage is a period of cell division in the absence of growth, whereas oogenesis is a period of growth without replication or division. The images are of a human zygote, and an eight-cell stage embryo.

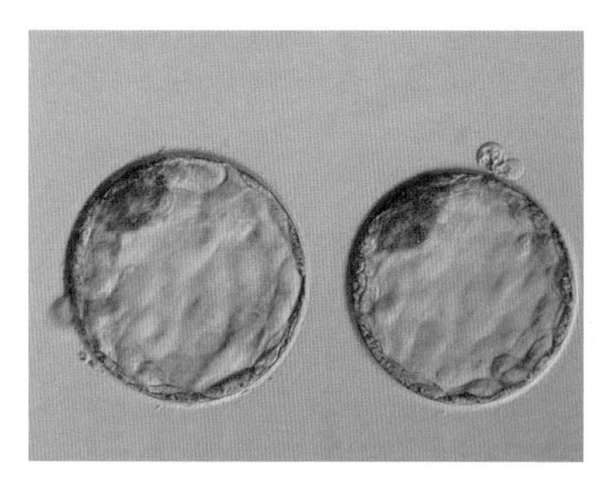

Figure 7.22 Two human blastocysts produced in an in vitro laboratory. Today two healthy young twins (courtesy of Centro Fecondazione Assistita, Naples).

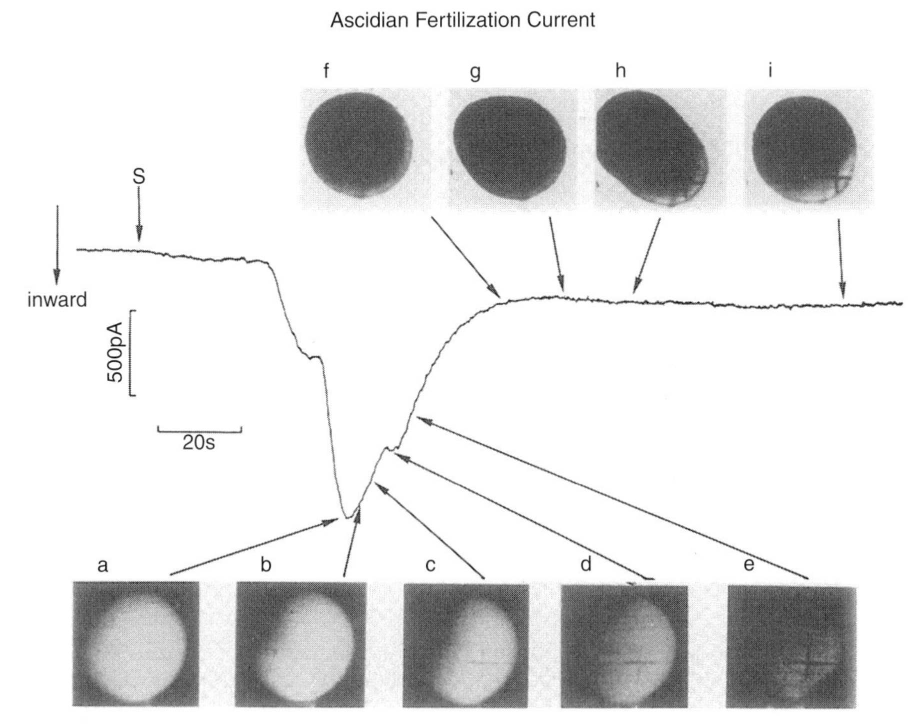

Figure 5.12 The activation current in the ascidian oocyte is followed by the release of Ca^{2+} from intracellular stores and a surface contraction that traverses the oocyte from the animal pole to the opposite hemisphere (from Brownlee and Dale 1990).

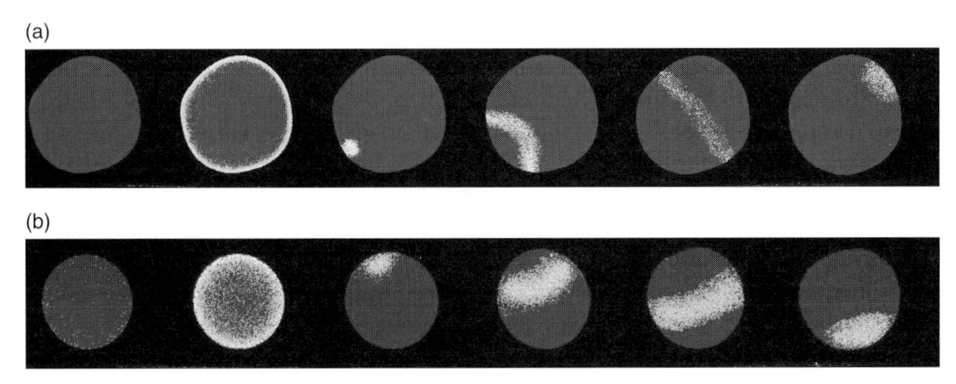

Figure 5.13 The cortical flash and subsequent calcium wave in **(a)** an oocyte of the starfish *Astropecten auranciacus* (180 μm) fertilized 70 minutes after the addition of 1-MA and **(b)** an oocyte of the sea urchin *Paracentrotus lividus* (80 μm). The cortical flash is thought to be due to entry of Ca^{2+} through ion channels in the oocyte plasma membrane from the surrounding medium, while the wave is caused by the release of Ca^{2+} from intracellular stores. The calcium wave traverses the sea urchin oocyte in 28 seconds, but owing to the large size of the starfish oocyte, it takes two minutes in the latter (courtesy of Dr Nunzia Limatola and Dr Luigia Santella of the Stazione Zoologica).

Activation of the Ascidian Oocyte

Ascidian, or sea squirts, are urochordates and have been used as a model for developmental biology since the days of E. Conklin in 1905. Their oocytes offer many advantages over other species and were selected for study by my laboratory in Naples from 1980 to 2000. First, they are easy to extract manually from their extracellular coats, exposing a nude

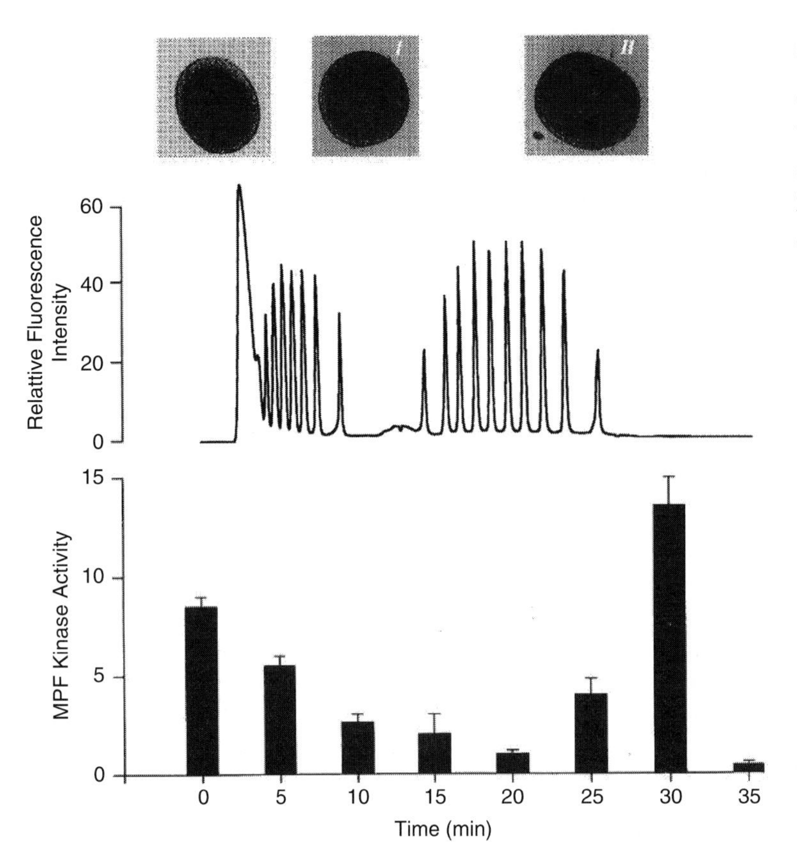

Figure 5.14 The relationship between intracellular calcium, MPF activity, cortical contraction and polar body extrusion in *Ciona intestinalis* oocytes at fertilization and progression through meiosis. The top panel shows the cortical contraction and the extrusion of the first and second polar bodies. The middle trace shows the calcium oscillations while the lower bar chart shows H1 kinase activity (from Russo et al. 1996).

plasma membrane that is extremely easy to patch and whole-cell clamp, which allows simultaneous electrical recordings and microinjection. Conklin showed that sperm entered the ascidian oocyte at the vegetal pole. We have confirmed this using homologous sperm in *Phallusia mammillata* and *Ciona intestinalis* and also heterologous sperm from the oyster *Ostrea edulis;* however, there are conflicting reports that the sperm enters the animal pole. Despite this controversy, it is clear that the calcium wave and subsequent surface contraction initiates at the animal pole and traverse the oocyte to the vegetal pole. The surface contraction follows and depends on intracellular calcium release but is independent of the fertilization current. Ascidian oocytes at spawning are blocked in meiosis 1, in contrast to vertebrate oocytes that are blocked at meiosis 2, giving the opportunity to study the control of meiosis through both meiotic divisions. Finally, in contrast to many vertebrate species, one animal provides sufficient numbers of oocytes for biochemical studies. It was possible, therefore, for the first time in an animal species, to correlate physiological, structural and biochemical events at activation (Figure 5.14).

The first calcium transient at activation can be detected about 20 seconds after insemination, shortly after the fertilization current, and increases intracellular calcium to 10 μM.When this calcium transient and the fertilization current are complete, the surface contraction starts in the same direction as the calcium wave. A region of elevated calcium persists at the animal pole. Locally applying calcium ionophore A23187 to the oocyte surface induces the contraction wave, but increasing intracellular calcium either by application of the ionophore or by stimulating voltage-gated calcium channels did not result in a surface contraction. Ascidian oocytes have both T- and L-type calcium channels, and considering the density and kinetics of the L-type calcium channels, it is highly probable that they play a role in regulating cytoplasmic calcium.

The large single transient peak is followed by several smaller oscillations that last for five minutes (phase 1). About nine minutes from insemination, after phase 1 was terminated, the first polar body was extruded. The intracellular calcium then stayed at baseline for five minutes (phase 2) when a second series of calcium oscillations started that lasts eleven minutes (phase 3) and terminates at the extrusion of the second polar body. Phases 1 and 3 can be inhibited by microinjecting heparin into the cell, but not the initial transient, showing that the late phase oscillations are mediated solely by IP_3, while the first peak is, in part at least, generated by ryanodine receptor channels (RYR). The phase 1 and phase 3 oscillations are essential for meiosis since blocking them with heparin inhibits polar body formation. Simultaneous, recording of membrane currents and intracellular calcium showed that the 1–2nA inward current correlated temporarily with the first activation peak of calcium. Phase 1 and phase 3 transients also differ since only the latter occur simultaneously with mirror image plasma membrane currents and an increase in HI kinase activity while the phase 1 oscillations do not. Second, the time intervals between the phase 3 transients is longer than the time interval between the phase 1 oscillations suggesting the two groups of calcium oscillations and the modulation of meiosis 1 and meiosis 2 may be spatially distinct.

The two main components of MPF are a catalytic sub-unit called Cdc2 kinase and a regulatory sub-unit called Cyclin B. Cdc2 activity is regulated by phosphorylation and de-phosphorylation on specific residues as well as physical binding with cyclins. Cyclin B is regulated by its synthesis and degradation and possibly by phosphorylation. MPF activity can be shown by measuring the ability of Cdc2 to phosphorylate the specific substrate histone H1 in vitro. Histone HI kinase activity in ascidian oocytes was high at metaphase 1 and decreased within five minutes of insemination, reaching a minimum at the phase 2 oscillations corresponding to telophase 1. During phase 3, histone H1 kinase activity increased again and then decayed again during telophase11. Oocytes preloaded with BAPTA, to sequester intracellular calcium, did not generate calcium transients; nonetheless, the HI kinase activity decreased five minutes after insemination and remained low for 30 minutes suggesting that the inactivation of MPF at meiosis 1 is independent of the calcium transients. However, these oocytes did not extrude the first polar body, suggesting calcium is required for completion of meiosis 1 and that there are a number of signals involved in meiosis 1 progression. ADPr probably plays a role in completion of, meiosis 1. In fact, nicotinamide, an inhibitor of nicotinamide nucleotide metabolism, blocks the initial calcium peak but enhances calcium influx by opening the ADPr-gated channel and the fertilization channel, and blocks MPF inactivation, possibly by maintaining high levels of Cdc2 activity.

In contrast, injection of BAPTA during the phase 2 oscillations suppressed the phase 3 calcium currents inhibiting the increase in H1 kinase activity and second polar body extrusion, showing that progression through meiosis 2 is calcium dependent. This is consistent with the calcium-dependent progression through metaphase 2 in mouse oocytes. This work on the ascidian oocyte was the first to show that the mechanism of progression through metaphase 1 is different to that for progression through metaphase 2. MAPK activity, a component of CSF that is thought to maintain the MPF activity high in meiotic arrest, is high in unfertilized ascidian oocytes, peaks at five minutes after insemination and then decreases at telophase 1 without any subsequent increase.

The first series of waves arise from a pacemaker at the sperm entry site that are related to contraction of the surface and ooplasmic segregation, and appear to depend on Rho ATPase activation. The second series of waves arise from the sperm centrosome and are generated from the equatorial region of the oocyte. MPF upregulates calcium wave production by modulating IP_3 levels, while MAPK seems to be necessary for the calcium oscillations in some ascidians, but is only required for the second oscillations in others. In summary, MAPK maintains the meiotic state, MPF promotes calcium oscillations during meiosis and then the calcium oscillations down regulate the MAPK to allow exit from meiosis.

Activation in Teleost Oocytes

In the teleost fish *Oryzias latipes*, oocytes blocked at metaphase 2, undergo a single calcium transient that starts at the site of sperm entry at the animal pole and traverses the oocyte at 10–15 μm/sec through the peripheral cytoplasm where the yolk is absent. The dynamics are a little different in the zebrafish (*Danio rerio*) where, in addition to the peripheral wave, a slower wave passes through the centre of the oocyte, and this is followed by several irregular smaller waves. In zebrafish oocytes, the SFK member Fyn is concentrated at the animal pole and appears to increase PLCγ that triggers an IP_3-dependent calcium wave in the cortex. In fish, the calcium waves are generated by both IP_3-

and non-IP_3-dependent pathways; however, fish oocytes may be activated in the absence of calcium waves being generated.

Activation in Amphibian Oocytes

Frog oocytes are also arrested at metaphase 2 and generate a single calcium wave that starts at the point of sperm entry and travels to the vegetal pole in 5–20 μm/sec by an IP_3-mediated mechanism. In *Xenopus,* fertilization appears to trigger a transmembrane oolemmal protein that triggers a non-SH2 mediated activation of a Src-type SFK to increase PLCγ activity and IP_3 levels. In response to the calcium wave in amphibian oocytes, calcium sensitive regulators such as calcineurin and calcium/calmodulin-dependent kinase 2 cause a decline in both MPF and MAPK.

The newt *Cynops* is physiologically polyspermic, with up to 10 spermatozoa normally entering the cytoplasm. The first sperm that enters triggers a Ca^{2+} spike, apparently due to the effect of a protease on the oocyte surface, shortly after a calcium wave is generated from the sperm entry point and traverses the oocyte at 5 μm/sec. Supernumerary sperm also trigger partially propagating waves so that collectively the Ca^{2+} levels remain high for some time. The calcium waves in *Cynops* are triggered by the sperm factor citrate synthase which seems to function by way of an IP_3 mechanism. The citrate synthase may function via PLCγ or alternatively it could inversely catalyse the cleavage of citrate into acetyl-CoA and oxaloacetate. Whatever the precise trigger, the calcium wave induces the inactivation of MPF and MAPK, and cell cycle progression is induced.

Activation in Mammalian Oocytes

In mammals, fertilization triggers a series of calcium oscillations in the oocyte with each transient lasting from one to two minutes. The oscillations continue until pronuclear formation, which could be up to 20 hours after fertilization. Individual transients occur every three minutes in mice and every 50 minutes in cows. The initial Ca^{2+} increase in the mouse spreads through the ooplasm at 20 μm/sec starting from the area of gamete fusion. As oscillations progress, they speed up and appear to originate from a pacemaker located distant to the sperm entry site, similar to the situation in the ascidians. The calcium waves appear to be due to IP_3-mediated calcium release from the endoplasmic reticulum. In addition to the internal stores, mammalian oocytes require external Ca^{2+} to maintain the oscillations and also for emission of the second polar body. In pigs, such influx may depend on store-operated calcium entry (SOCE). In mammals, the calcium oscillations activate CAMKII, which then drives cell cycle progression and meiotic resumption by the proteolytic degradation of the cyclin BI subunit of MPF and degradation of MAPK. The calcium signals also induce the release of zinc by an exocytotic process that may help in meiosis resumption.

Activation in the Protostomes

The protostomes include annelids, flatworms, molluscs and arthropods. Here, if the blastomeres are physically separated in a four-cell embryo, each blastomere will produce a quarter of the gastrula and larva. Thus the fate of blastomeres is fixed very early in development, and cleavage is termed determinate. The cleavage plane in protostome embryos is oblique, resulting in a spiral cleavage pattern. Furthermore in protostomes, the mouth forms from the blastopore and coelom formation occurs by a process called schizocoely, where a split forms in the mesodermal mass, forming a cavity that enlarges and becomes the coelom.

Activation in Annelid Oocytes

Annelid oocytes are arrested at metaphase 1. At fertilization in *Chaetopterus* a non-propagated calcium wave is generated followed by a series of propagating calcium waves. In *Pseudopotamilla* the first calcium increase is restricted to a cortical region, which causes an extrusion of the cortex to engulf the fertilizing sperm. This is followed by a cortical flash and subsequently a larger global rise that is responsible for meiotic resumption. In annelids, voltage-gated channels are responsible for the calcium influx, while an IP_3-mediated mechanism leads to calcium release from internal stores. Echiuran worms, such as *Urechis,* are arrested at prophase 1 and generate a non-wave-like increase in peripheral calcium by external calcium flowing into the oocyte, possibly supplemented by intracellular calcium release by an IP_3-dependent mechanism.

Activation in Molluscan Oocytes

Molluscan oocytes are fertilized at prophase 1 or metaphase 1, depending on the species. They generate single or repetitive calcium waves that last 5–30 minutes. The initial Ca^{2+} increase is a cortical flash caused by influx through voltage-gated channels, followed by an elevated plateau in some species or a series of Ca^{2+} oscillations in others, which is released from IP_3-sensitive stores.

Activation in Nemertean Oocytes

Marine nemertean oocytes are arrested at metaphase 1. Fertilization triggers a cortical flash arising from external Ca^{2+} influx, followed by multiple calcium waves that travel at 10–15 μm/sec every 2–8 minutes until polar body formation. The initial waves start at the sperm entry site while the subsequent waves are generated from near the vegetal pole.

Activation in Nematode Oocytes

In the round worm, *Caenorhabditis elegans* fertilization occurs between GVBD and meiosis 1. The oocytes are ovulated from the gonad into a spermatheca, and the first side of the egg to enter the spermatheca is where the calcium wave is initiated. The calcium concentration in the cytoplasm increases to 250–500 nM.

Activation in Arthropod Oocytes

In many arthropods the oocyte is pre-activated before encountering the sperm. In insects for example, activation is triggered by mechanical or osmotic forces, as the oocytes pass through the female reproductive tract and such mechanisms appear to function through calcium signalling. *Drosophila* fertilization has been studied the most, and here it is thought that oviductal fluid causes osmotic swelling to the oocyte posterior which in term activates mechano-sensitive calcium channels leading to an increase in intracellular Ca^{2+}. The single Ca^{2+} wave propagates in an actin dependent manner leading to resumption of the cell cycle and eventually translation of specific stored mRNAs such as *bicoid*. In oocytes of the marine shrimp *Sicyonia*, which are arrested at metaphase 2, activation is triggered by exposure to the Mg^{2+} ions in seawater that then elicits a calcium wave. There is apparently a second calcium wave induced at fertilization.

Activation in Cnidarian Oocytes

In the Cnidaria (jellyfish), the calcium wave is a single transient that lasts 2–10 minutes and travels at 4–6 μm/sec (fast wave), starting from the animal pole, the site of sperm entry. After propagating across the oocyte the wave leads to a dephosphorylation of MAPK (mitogen activated protein kinase) and cell-cycle progression through the G_1/S transition. The same effect may be mimicked by injecting IP_3 into the oocyte, whereas cADP (cyclic AD-ribose) that stimulates the ryanodine receptor pathway has little effect. High K^+ solutions will generate a cortical flash in jellyfish oocytes although to date it has not been shown to occur at fertilization.

pH and Activation

One of the first signs of oocyte activation in the sea urchin oocyte is the acidification of the surrounding seawater due to Na^+/H^+ exchange. The oocyte cytoplasm as a result increases in pH by 0.4 units. Blocking this pH_i increase by the drug amiloride or removing external Na^+ does not affect Ca^{2+}-dependent events such as the cortical reaction, however it does stop all downstream activation events such as pronuclear movement and DNA synthesis. Artificially increasing pH_i leads to the activation of these later events without the earlier Ca^{2+} dependent events, showing that activation is regulated by dual ionic signals that are temporally separated (Figure 5.15).

Structural Changes at Activation

Simultaneously or shortly after the plasma membrane and calcium events, oocytes undergo a massive reorganization of their surface manifest as a cortical contraction or the exocytosis of cortical granules. Cortical granules are found in the oocytes of many animals, for example echinoderms, some mammals, amphibians, annelids, fish and crustaceans, although there is tremendous diversity in their form and size. In mammals, sea urchins and starfish they are found immediately below the plasma membrane in a single layer, whereas in some ophiuroid echinoderms (brittle stars) they are found several layers deep. A similar massive cortical reaction is to be found in the oocytes of the shrimp *Penaeus* and the cnidarian *Bunodosoma*. These special organelles originate as vesicles in the Golgi complex and in sea urchins accumulate throughout the oocyte cytoplasm until germinal vesicle breakdown (GVBD) when they move to the periphery. In starfish oocytes, as in mice, the cortical granules are directed to the periphery as soon as they are synthesized, gradually increasing the density of these granules throughout oogenesis. There appears to be different populations of cortical granules in many species studied to date based on their electron density. In fish, there are two types of granule, electron dense and electron lucent. In the amphibian, *Xenopus laevis*, two different types of granules are distributed topographically. One type found at the

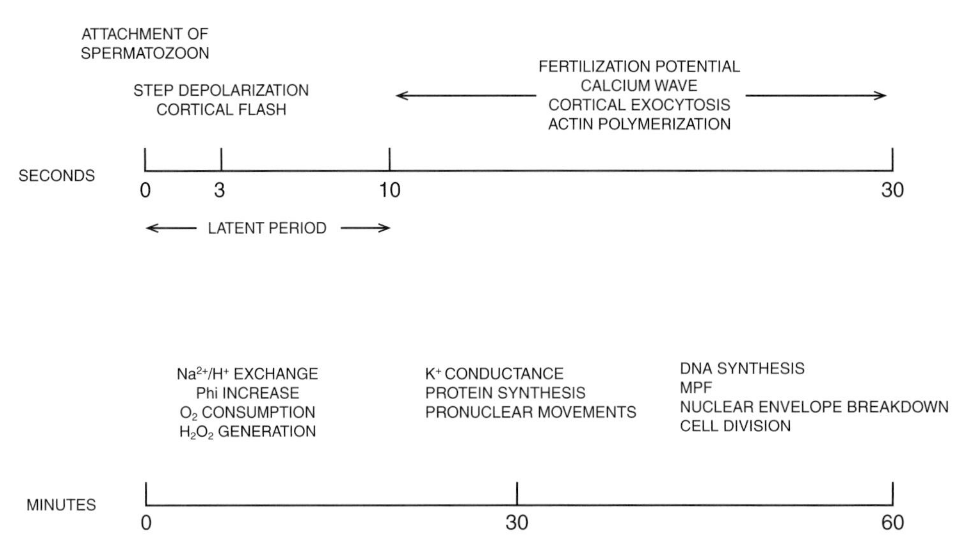

Figure 5.15 The sequence of changes in the sea urchin oocyte at fertilization. The two major cytoplasmic ion messengers, Ca^{2+} and pH_i, are temporally separated and regulate different activation events.

animal pole has a homogeneous matrix of moderate electron density; the other found at the vegetal hemisphere contains loose flocculent material. In the lobster *Homarus americanus*, there are four types of cortical granules based on their ultrastructural appearance with various electron densities. In sea urchins, different populations of cortical granules have been distinguished by biochemical and immune cytochemical methods. Even in oocytes that lack cortical granules, such as those of the ascidian, there appears to be some sort of exocytosis of cortical material at activation and, in some species at least, the perivitelline space expands.

The Cortical Reaction in Sea Urchins

Most of our information on cortical granule exocytosis has come from studies on the sea urchin oocyte. There are from15,000 to 20,000 cortical granules per oocyte, which are synthesized during oogenesis. They are bound by membranes that measure about 1 µm in diameter, and their contents often look star-or spiral-shaped. The granules appear to be interconnected by thin filaments and are firmly attached to the oocyte surface. A tangential section through an oocyte clearly shows their homogeneous distribution just below the cell surface (Figure 5.16). Exocytosis starts some 10 seconds after the fertilizing spermatozoon has attached to the vitelline coat, and this delay correlates with the period from the electrical step event to the fertilization current and is called the latent period. The granules in the immediate vicinity of the spermatozoon are the first to break down, and a wave of

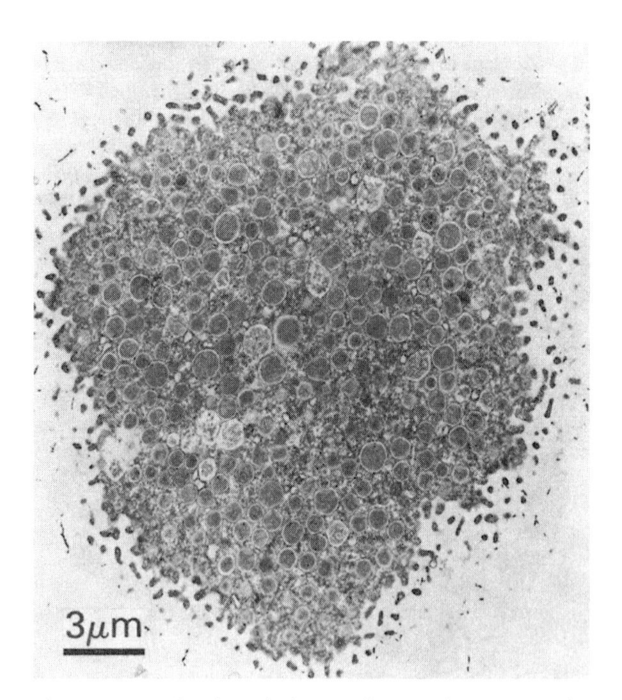

Figure 5.16 A slice through the top of a sea urchin oocyte at the transmission electron microscope showing the homogenous distribution of cortical granules (from Dale 1983).

(a)

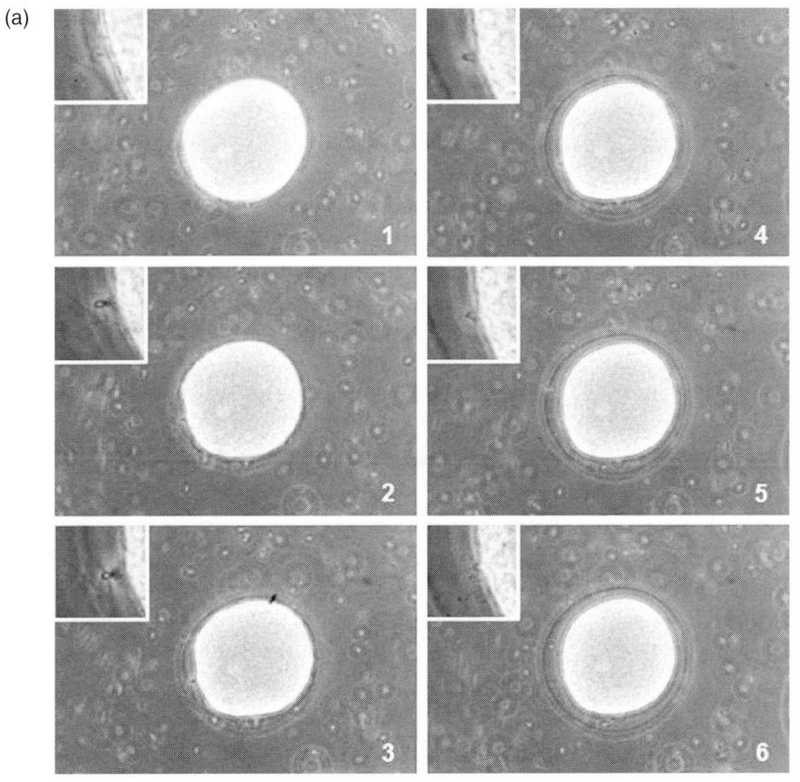

Figure 5.17 **(a)** A series of light micrographs taken at five-second intervals of a sea urchin *Paracentrotus lividus* oocyte showing the entry of the successful spermatozoon (to the left) and at higher magnification in the insets. Note the flattening or concavity at the sperm-attachment site and the elevation of the fertilization membrane in Frame 2 that spreads around the oocyte in a wave, taking about 20 seconds to form a global protective covering. **(b)** A diagram showing cortical exocytosis and resulting elevation of the fertilization membrane in the sea urchin oocyte (from Dale 1983). (A black-and-white version of this figure will appear in some formats. For the colour version, please refer to the plate section.)

(b)

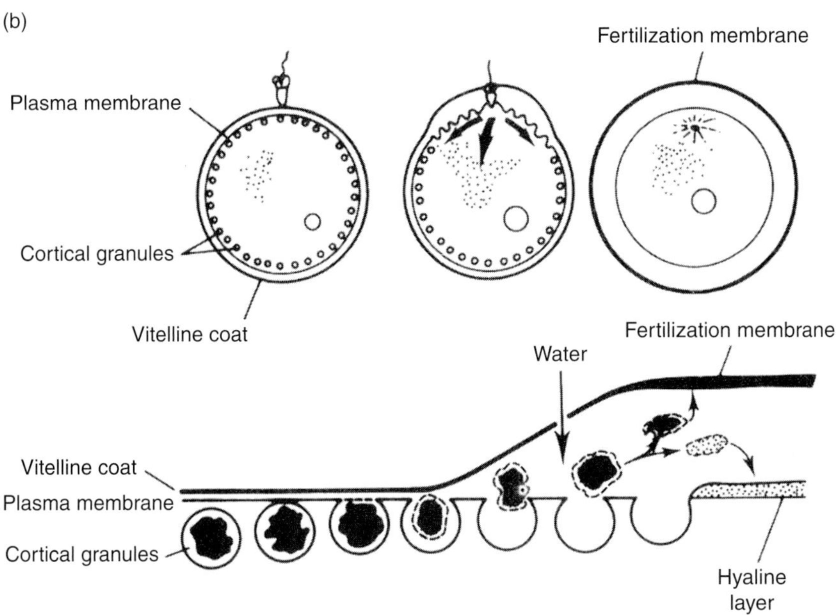

exocytosis then spreads slowly around the oocyte surface (Figure 5.17). The wave of cortical granule exocytosis is self-propagating. Mucopolysaccharides, released from the granules into the perivitelline space, cause a rapid influx of water, which distends the vitelline coat, lifting it by hydrostatic pressure 10–20 µm away from the oocyte surface. This wavelike progression of cortical granule exocytosis and the concomitant elevation of the vitelline coat take approximately 20 seconds to spread around the oocyte surface (Figure 5.17). Over the next five minutes, the perivitelline space continues to increase in volume, and the vitelline coat hardens and thickens, becoming the familiar fertilization membrane. Although it is not clear how the cortical reaction is initiated, nor how it is propagated around the oocyte, an important factor is the increase in cytoplasmic Ca^{2+}. Cortical granule exocytosis may be induced by injecting Ca^{2+} into the cytoplasm, and it can be prevented using Ca^{2+}-chelating agents (molecules which bind free Ca^{2+}). Furthermore, oocytes exposed to ionophores (molecules which facilitate the passage of ions across membranes) such as amphotericin B or A23187 undergo the cortical reaction along with several other activation events. Using isolated preparations of sea urchin cortices it has been shown that 9–18 μM of free Ca^{2+} is sufficient to induce the exocytosis of cortical granules and this is in the same order as measured in oocytes in vivo during the fertilization wave.

In sea urchin oocytes, cortical granules contain calcium, serine protease, peroxidase, hyaline protein, beta-glucuronidase, beta 1,3-gluconase, sulphated mucopolysaccharides, the soft fertilization envelope proteins SFE1 and SFE9, proteoliaisin, rendezvin and ovoperoxidase. SFE1 and SFE9, and proteoliaisin are proteins rich in low-density lipoprotein receptor type A (LDLrA) repeats involved in protein interaction, while rendezvin is enriched in Complement C1r/C1s, Uegf, Bmp1 (CUB) domains, which are also involved in protein interactions. During activation, the granules break open, releasing their contents into the perivitelline space (the gap between the oocyte plasma membrane and the vitelline coat). There are two immediate consequences of cortical granule exocytosis: the perivitelline space first increases in volume, and the vitelline coat is then transformed into a thick, hard protective structure, the fertilization membrane (Figure 5.17). The fertilization membrane is a permeable structure that allows exchange with the surrounding seawater. In starfish, the fertilization membrane allows the diffusion of molecules up to 2,000 kDa, while that of the sea urchin is less permeable allowing the diffusion of molecules of up to 40 kDa.

Proteolysis, transamidation, hydrogen peroxide synthesis and peroxidase-dependent dityrosine cross-linking are the four major enzymatic activities required for the assembly of the fertilization membrane. Cortical granule serine protease is inactive when it is contained in the cortical granules due to the low pH (6.5), while following exocytosis the protease is activated by autocatalysis following exposure to seawater (pH 8.2). This serine protease limits the activity of ovoperoxidase and increase the activity of Beta 1,3 glucanase while it is also targets a protein (p160) that links the vitelline layer to the plasma membrane, allowing its separation at fertilization. Transamidation is mediated by transglutaminases that cross-link glutamine and lysine. Two transglutaminases, extracellular transglutaminase and nuclear transglutaminase, have been found in *Strongylcentrotus purpuratus*. Hydrogen peroxide is synthesized at fertilization for ovoperoxidase cross-linking activity

Finally, it should be pointed out that cortical granule exocytosis is essentially a process of membrane fusion, and each granule fuses with the inner aspect of the plasma membrane. In order to fuse, the granules must come into close contact with the oocyte surface, and the cytoskeleton is involved in this movement. The oocyte plasma membrane becomes a mosaic structure made up of the original membrane and the incorporated patches of cortical granules. The physiological characteristics of these inserted patches of new membrane are not known, nor is it known whether they are essential for embryogenesis, however, they do cause a 100 per cent increase in the surface area of the oocyte. There are clear structural modifications resulting from the cortical reaction. The vitelline coat of the unfertilized oocyte is a glycoprotein structure approximately 15 nm thick. Immediately following elevation, it remains thin and elastic and can easily be digested by proteolytic enzymes. During the next five minutes it becomes thicker, reaching approximately 90 nm, and much harder. The fertilization membrane is a laminar structure, and it is thought that the crystalline component of the cortical granules attaches to the inner aspect of the vitelline coat thereby thickening it. A trypsin-like protease also released from the cortical granules plays an important role in this thickening and hardening – if this protease is specifically inhibited, the fertilization membrane fails to harden. A further component of the granules is a calcium-binding glycoprotein, which tightly adheres to the oocyte forming the hyaline layer. In actual fact, a very thin hyaline layer may be present before fertilization (also of cortical granule origin), which then thickens as a result of the cortical reaction. Peroxidase is also released from the cortical granules and contributes to hardening of the fertilization membrane by cross-linking tyrosine residues aided by hydrogen peroxide. Peroxide synthesis occurs as a burst at fertilization and accounts for two-thirds of the oxygen uptake by the oocyte during the first 15 minutes after fertilization.

The Cortical Reaction in Mammals

In mammalian oocytes, the cortical granules are smaller, ranging in size from 0.2 to 0.6 μm, and are synthesized in the early stages of oocyte growth, the exact time being species specific. In rat and mouse, they first appear in the unilaminar follicle, while in humans, monkeys and the hamster they first appear in multilayered follicles. At first, small vesicles are formed from the Golgi that then migrate to the sub cortical area and coalesce to form mature cortical granules. The production of cortical granules is a continuous process up until the time of ovulation. Granule migration depends on cytoskeletal elements in all animals including the sea urchin. Rab proteins constitute the largest family of monomeric small GTPases. These proteins function as molecular switches between active (GTP-bound) and inactive (GDP- bound) conformations. During their active state they cooperate with downstream 'effector' proteins, which are involved in different cellular activities such as vesicle formation, motility, the movement of vesicles and organelles along cytoskeletal elements. In mouse oocytes the cortical granules translocate along the actin network regulated by Rab27a, which is a marker for cortical granules, using two pathways. In the first, myosin Va transports the granules along the actin, while in the second the granules attach to Rab11a vesicles that move to the plasma membrane.

In mice and hamster oocytes, the granules are evenly distributed through the cortex, except for the area over the spindle that is devoid of granules. It has been calculated that upto 24 per cent of the surface area around the animal pole, the site of the meiotic spindle, is devoid of cortical granules. This is not the case for all other mammals including the human, where the granules are located all over the cortex. In the mouse the number of cortical granules decreases from 7,400 to 4,100 after extrusion of the polar bodies, suggesting a number are exocytosed in this area before fertilization, excluding the possibility of sperm penetration in the area overlying the female meiotic plate. The pre-fertilization release of cortical granules in some mammals may serve to condition the zona pellucida for subsequent interaction with the spermatozoa.

Two classes of proteins, v and t SNARES (soluble NSF attachment protein receptors) have been shown to play a role in cortical granule docking and exocytosis. Alpha-SNAP (N-ethylmaleimide-sensitive factor attachment protein alpha) and NSF (N-ethylmaleimide sensitive factor) are also indicated in the cortical reaction in mouse.

It has been estimated that there are between 4 and 14 proteins in the cortical granules of mammalian oocytes. Various techniques, including lectin staining has shown the granules to be rich in carbohydrates where the carbohydrate moieties on glycosylated constituents are complex. There are several proteinases present in mammalian cortical granules, one being a trypsin like molecule. Tissue-type plasminogen activator (tPA), a serine proteinase and ZP2 proteinase, involved in ZP2 proteolysis, which converts ZP2 (120 kDa) to ZP2f (90 kDa) have also been found. An ovoperoxidase, involved in catalysing the cross-linking of tyrosines in the zona pellucida to harden it, has been demonstrated to be in mammalian cortical granules. Calreticulin, a chaperone protein responsible for glycoproteins, N-acetylglucosaminidase, a glucosidase, p32, p75 have all been recognized in mouse cortical granules by specific antibodies though their roles are not clear. Ovastacin, a protein of the metalloproteinase family, similar to hatching enzyme, is also found in cortical granules. In the human and other mammals, two populations of cortical granules can be distinguished based on their electron density, one has an electron dense core, the other has fluffy or granular contents, which could reflect different stages in granule maturity, different granules or different stages of exocytosis.

The cortical reaction in the mammalian oocyte elicits the zona reaction, changing the characteristics of the zona pellucida and the oocyte plasma membrane, which now becomes a mosaic of cortical granule membrane and the original plasma membrane. The alteration to the zona pellucida in the mouse, involving proteinases or glycosidases, cause the hydrolysis of ZP3 receptors changing its role from a sperm receptor to a protective coat.

The Cortical Reaction in Fish

Oocytes of the teleost fish *Oryzias* have cortical alveoli of 10–40 μm, except for the area around the micropyle,

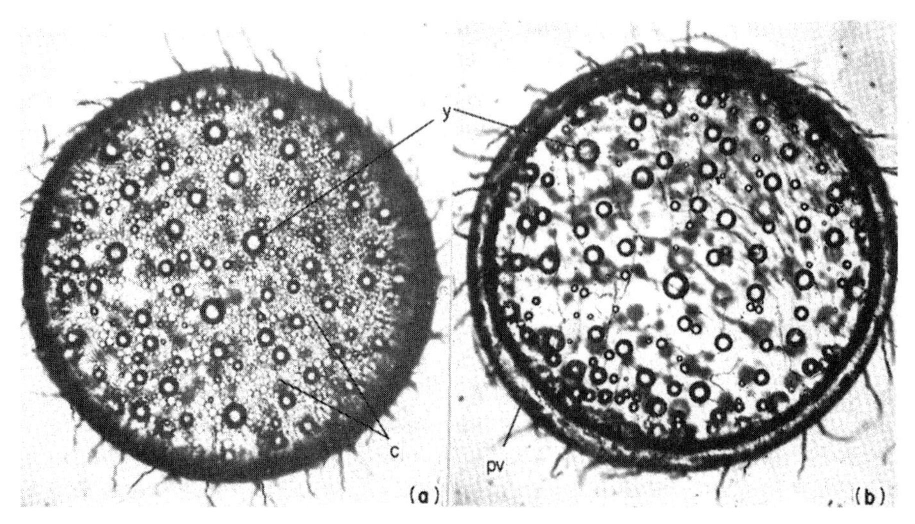

Figure 5.18 The unfertilized oocyte of the fish *Oryzias* is 1.2 mm in diameter and contains large yolk drops and smaller cortical alveoli (left frame). After fertilization, the cortical alveoli break down and the perivitelline spave (pv) increases in volume (right frame) (from Dale 1983).

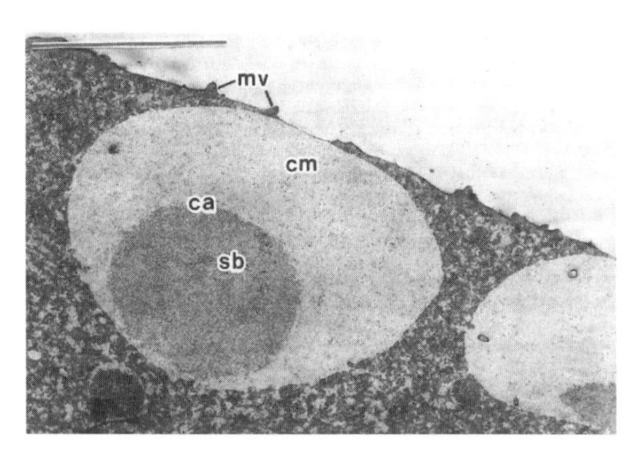

Figure 5.19 Cortical alveoli in the cortex of the oocyte of the teleost fish *Oryzias latipes*. ca – cortical alveolus; cm – colloidal material; mv – microvilli, sb – spherical body. Bar = 10 µm (from Tarin and Cano 2000).

where there are smaller alveoli of 2–10 µm (Figure 5.18). The cortical alveoli appear before the initiation of vitellogenesis as small vesicles that move through the cortical layer as yolk formation proceeds. At the electron microscope, the alveolar content appears as a colloidal fluid containing a spherical body (Figure 5.19). The alveolae contain polysaccharides combined with a sialoglycoprotein of 15–100 kDa.

At fertilization, the cortical alveoli break down in a wave-like progression from the point of sperm entry, releasing their contents into the perivitelline space (Figure 5.18). A hyaline layer is formed, the chorion hardens, and water enters into the vitelline space, causing it to increase in volume. After fertilization, the chorion in fish undergoes a calcium-dependent hardening which involves a 10-fold increase in its mechanical strength. Hardening involves an oocyte exudate, which contains a proteolytic enzyme, and the mechanism involves a transglutaminase. Cortical exocytosis is triggered by the increase in cytosolic calcium at fertilization that may interrupt the organization of the microfilaments that separate the alveoli from the plasma membrane. Following exocytosis, the alveolar membrane fuses with the oocyte plasma membrane increasing its volume, however in contrast to the situation in the sea urchin, the perivitelline space swells not by distension of the chorion (vitelline coat) but by shrinkage of the oocyte surface. The small alveoli at the sperm entry site also exocytose at fertilization starting about nine seconds after sperm attachment. About five seconds later, flagellar movement of the spermatozoon ceases, which could indicate sperm–oocyte fusion.

The Cortical Reaction in Annelids

Not all annelids have cortical granules, for example oligochaete oocytes are devoid. Some spionids, sedentary tubicolous worms which include the commonly used *Chaetopterus*, in contrast have cortical alveoli. The cortical alveoli of the marine worm *Nereis* also break down in a wave-like fashion during activation and, similar to the situation in fish, the perivitelline space increases by oocyte shrinkage rather than by

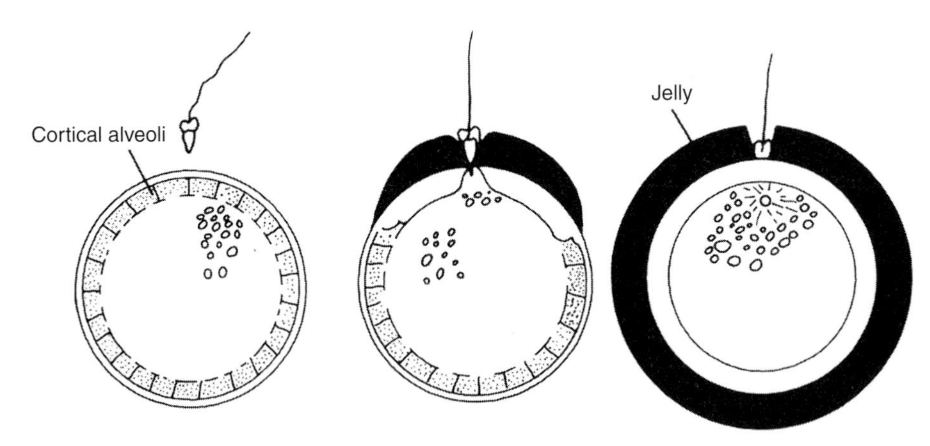

Figure 5.20 The cortical reaction in the marine worm *Nereis*. The alveoli breakdown in a wave like fashion, the oocyte shrinks and the material from the alveoli forms an external impervious jelly-like layer (from Austin 1965).

membrane elevation. A peculiar observation is that the alveolar contents extrude outside the oocyte surface layers and by hydration form a thick impervious jelly-like structure (Figure 5.20).

The Cortical Reaction in Amphibians

In anurans, cortical granule exocytosis and elevation of the vitelline coat is remarkably similar to that in the sea urchin. The wave of cortical exocytosis starts at the site of sperm entry, or site of pricking in the case of parthenogenetic activation, and spreads around the oocyte to the antipode. In these large oocytes it takes about one minute for the wave to cross the equator and 2.5 minutes to reach the opposite pole. Cortical granules in amphibians contain proteases and lectins, which transform the vitelline envelope into the fertilization membrane. In *Xenopus laevis* the contents of the cortical granules are deposited in the perivitelline space after exocytosis and diffuse through the vitelline membrane to the inner layer of jelly. A lectin in the exudate precipitates with a ligand in the jelly in a calcium dependent mechanism to form a fertilization layer. This occurs within four to eight minutes of fertilization. The lectin is a mixture of large glycoproteins (about 500–700 kDa) that specifically recognizes a terminal alpha-galactose of the ligand in the jelly layer. Urodeles lack cortical granules however some sort of cortical exocytosis occurs also here, since there is a deposition of macromolecules into the perivitelline space and changes in the vitelline envelope that cause the perivitelline space to increase three to four hours after fertilization. In the frog, the jelly layer swells immensely after fertilization and serves both for protection, as an attachment mechanism to submerge the oocyte and to keep the oocytes spaced for efficient metabolic turnover with the environment.

The Role of Cortical Reorganization

Cortical reorganization is a common feature of oocyte activation. Although mechanisms differ, the results are often similar. The sea urchin oocyte has been studied in great detail – other oocytes less so. By piecing together all of the information, some general conclusions may be drawn regarding the role of cortical reorganization in embryogenesis. First and foremost, embryos must be protected in some way. In oocytes which lack cortical granules, for example those of ascidians and insects, a protective structure, which does not alter much following activation, is laid down during oogenesis – i.e. it is preformed. In other animals a different strategy is employed. The oocyte has a relatively thin extracellular coat, which hardens after activation, catalysed by the cortical granule products. In either event, the embryo remains in a protective coat until hatching. A second extracellular structure produced as a result of cortical granule exocytosis is the hyaline layer, which serves to keep the dividing blastomeres of the embryo in close contact. The early compact mass of continually dividing cells in the embryo is therefore continually changing shape. Such movement would be hindered if the cells were attached to a rigid structure, so possibly for this reason the embryo is surrounded by the fluid-filled perivitelline space. This gap may also serve as micro-environment buffering the embryo from changes in the environment.

Reorganization of the plasma membrane appears to be related to the metabolic de-repression of the oocyte and occurs in oocytes both with and those without cortical granules. Certainly, this reorganization is dramatic and rapid in granule-containing oocytes and is attained without the participation of the cells' synthetic apparatus. Although we do not know the function of this mosaic plasma membrane, the resulting transient increase in surface area will facilitate the metabolic turnover of the activated oocyte.

Contractions of the Oocyte Surface and Cortical Actin

In many animals, the oocyte at activation undergoes a surface contraction, often in the same direction as the preceding calcium wave. For example, in anurans, cortical granule exocytosis is followed by a surface contraction, which is manifested by a combination of elongation of microvilli and smoothing of the oocyte surface. The contraction of the pigmented cortex in the animal hemisphere in anurans is called the cortical contraction and is due to an actin–myosin network in the surface regulated by Ca^{2+}. Oocytes from *Urodeles*, which do not have cortical granules, also undergo surface changes at activation manifested by a cyclical change in the length of microvilli that is thought to be due to changes in the actin cytoskeleton. In echinoderms there is a massive reorganization of the cortical actin system after fertilization that starts with the formation of the fertilization cone and traverses the oocyte as a wave in the same direction of the calcium wave.

Ascidian oocytes also do not have cortical granules, although they do appear to undergo exocytotic events at fertilization. In any case, they undergo a surface contraction that traverses the oocyte from the animal pole to the vegetal pole that is driven by an actomyosin mechanism. Surface contractions in oocytes at activation cause the redistribution of organelles and other cytoplasmic components in the zygote setting the scene for the segregation of components into blastomeres and the polarized development of the early embryo and the establishment of polarity.

Cortical Actin in Sea Urchin Oocytes

The plasma membrane of sea urchin oocytes is organized in short microvilli that lack a core of actin microfilaments, whereas in amphibians, molluscs and mammals, the microvilli are longer and have a microfilamentous core. In sea urchins, cortical actin not only causes surface contractions but also plays a fundamental role in regulating oocyte physiology. For example, altering the structural organization of the actin cytoskeleton in the cortex significantly alters the amplitude and duration of the Ca^{2+}-induced cortical flash and the shape of the global Ca^{2+} wave. The dynamics of cortical granule exocytosis and the elevation of the fertilization envelope can be altered or even repressed, even in the presence of the Ca^{2+} wave, if the cytoskeletal actin in the cortex is perturbed. In fact, work from the group of Luigia Santella at the Stazione Zoologica suggests that the actin cytoskeleton is also a Ca^{2+} store that is mobilized at fertilization. It has been known for decades that actin filaments are involved in sperm incorporation. In sea urchin oocytes, actin is polymerized at the sperm-binding site to form a fertilization cone, a specialized structure to facilitate sperm incorporation. The internalization of the sperm occurs concomitantly with the translocation of actin fibres towards the centre of the fertilized oocytes.

The Actin Cytoskeleton in Starfish Oocytes

Starfish oocytes extracted from the gonad during the breeding season contain an almost homogenous population of immature oocytes blocked at prophase 1 of the meiotic cycle showing the large germinal vesicle nucleus. Maturation is induced by adding the hormone 1-methyladenine (1-MA) to the seawater, which induces a rapid reorganization of the actin cytoskeleton in the oocyte surface, accompanied by a Ca^{2+} increase in the cortex and in the GV. The germinal vesicle then breaks down (GVBD), and the nucleoplasm mixes with the cytoplasm, an essential process for cortical maturation and oocyte activation. 1-MA acts on the surface of the oocyte, interacting with G-protein-coupled receptors and activating MPF. GVDB in the starfish oocyte requires at least an eight-minute exposure to a concentration of 1mM of 1-MA.

The optimal period for fertilizing starfish eggs is between GVBD and the extrusion of the first polar body. The average diameter of fully grown oocytes is 250 μm in the Mediterranean species *Astropecten aranciacus*, and 180 μm in the Japanese species *Asterina pectinifera* while the GV is 80–100 μm in diameter and contains a single nucleolus. The GV is anchored to the animal cortex by microtubular structures and possibly actin filaments.

Within two minutes of adding 1-MA, the actin cytoskeleton reorganizes, manifested by the retraction of microvilli, alignment of the cortical granules with their longer axis perpendicular to the plasma membrane, and the formation of surface spikes that contain bundles of actin filaments (Figure 5.21). These structural modifications, generated by the rapid, reversible polymerization of actin, are reflected in changes of the mechanical properties of the oocytes, such as a decrease in the stiffness of the endoplasm. There are also cyclic changes in surface stiffness during the emission of the polar bodies, which is linked to the reorganization of the actin filaments forming a contractile ring at the animal pole. The actin network also appears to be essential for the delivery of chromosomes to the microtubule spindle during the first nuclear division of meiosis. The mechanical changes in the maturing oocyte, which occur at the level of the cortex and the inner cytoplasm, can be disrupted by cytochalasin, which disrupts F-actin-based structures. Starfish oocytes contain two populations of cortical actin filaments: spikes and non-spikes F-actin.

In immature starfish oocytes exposed to 1-MA, there is a transient increase in intracellular calcium, which originates from the nucleus and cytoplasm and is propagated in a mechanism distinct to the fertilization calcium wave. The 1-MA-induced Ca^{2+} increase always originates at the vegetal hemisphere of the oocyte, although the entire oocyte surface is exposed to 1-MA, and it traverses the oocyte much faster, reaching the antipode in 20 seconds compared to the fertilization calcium wave that takes 80 seconds to reach the antipode. NAADP (nicotinic acid adenine dinucleotide phosphate) and cADPr (cyclic ADP ribose) do not appear to contribute to the 1- MA-induced Ca^{2+}, while also the role of IP_3 is debatable.

Dr Santella in Naples showed that mature starfish oocytes release more Ca^{2+} in response to a fixed amount of IP_3 than immature GV oocytes, showing that the intracellular Ca^{2+}-release mechanism also changes during meiotic maturation. This is not due to the amount of Ca^{2+} stored in the endoplasmic reticulum, which is the same in both stages, but due to sensitization of the calcium-release mechanism that occurs simultaneously with the structural reorganization of the actin cytoskeleton. In oocytes from many species, including mammals, the formation and disassembly of clusters of the endoplasmic reticulum is under the control of MPF. These clusters are responsible for the shape of the Ca^{2+}oscillations at fertilization. In starfish oocytes, the calcium-release mechanisms are activated and visible before the sperm head is even close to the oocyte.

Cortical Contraction in Ascidians

Giuseppina Ortolani in 1955 in Palermo was the first to demonstrate cortical contraction in ascidian oocytes, and this confirmed the work of E. Conklin in 1905 that cytoplasmic components were segregated to different parts of the oocyte cytoplasm shortly after

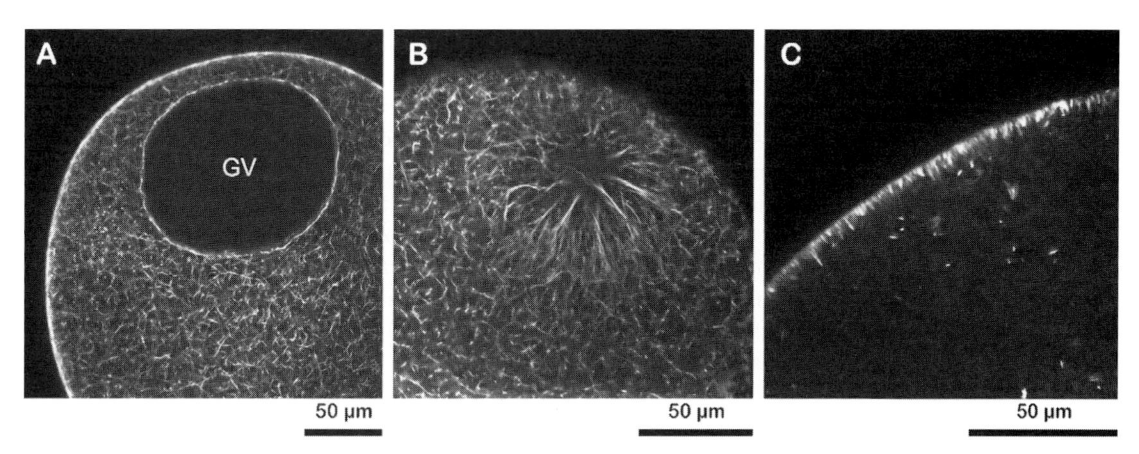

Figure 5.21 Changes in the actin cytoskeleton in the oocyte of the starfish *Astropecten aurantiacus* during maturation. **(a)** The GV stage showing a general dense distribution of F-actin in the cortex, **(b)** the radial arrangement of F-actin fibres over the surface of the germinal vescicle and **(c)** the perpendicular spikes of actin fibres reformed in the cortex of the mature oocyte (from Santella et al. 2015). (A black-and-white version of this figure will appear in some formats. For the colour version, please refer to the plate section.)

fertilization. The first indication of this cortical contraction is a restriction that appears at the animal pole that traverses the oocyte to the vegetal pole about one to two minutes after insemination and takes one to three minutes to pass from pole to pole. This wave is accompanied by a change in the surface microvilli that disappear at the animal pole and concentrate at the vegetal pole. Sub-cortical mitochondria and the cER mRNA are transported with this wave of contraction towards the vegetal pole and concentrate at this pole, forming a cytoplasmic cap which contains transparent cytoplasm. The cap disappears at the time of polar body expulsion some 8–10 minutes later. The fertilization contraction in ascidian oocytes occurs one to two minutes after the calcium wave and is driven by a cortical actomyosin mechanism probably mediated by Rho proteins.

Cortical Contraction in Mammals

Fertilization in mammals also induces rhythmic cortical cytoplasmic movements caused by contractions of the actomyosin cytoskeleton triggered by the calcium wave. Sperm entry in mammalian oocytes cause changes in the shape of the oocyte, mainly the formation of the fertilization cone and flattening of the zygote along the axis that bisects the fertilization cone. The sperm-induced cytoplasmic movements in the mouse last for four hours, until the formation of the pronuclei, and are synchronous with pulsations seen in the fertilization cone. These pulsations depend on the intracellular calcium waves and are actin dependent. One possibility is that the calcium activates kinases such as protein kinase C or Ca^{2+}/calmodulin-dependent kinase 11, which are known to regulate the cytoskeleton.

Aging and Cytoplasmic Maturity

In the life history of the oocyte, maturation sets in motion a chain of physiological and structural changes, giving rise to a cell unit geared to interact with the fertilizing spermatozoon at a particular moment in time. However, this condition is a transient one; should the oocyte not be fertilized, the maturation processes continues and the cell ages. Cytoplasmic aging is often independent of nuclear events, as first shown by Delage in 1901, and may lead to polyspermy, parthenogenesis, apoptosis, early extrusion of cortical granules, a decrease in MPF and MAPK, and changes in histone acetylation.

The phenomenon of aging of marine invertebrate oocytes is well documented. Oocytes may age in the ovary, following spontaneous ovulation, or after removal from the ovary; however, in the latter situation the process is considerably accelerated. Borei showed that the rate of oxygen consumption of sea urchin oocytes removed from the ovary rapidly and steadily declines, whereas overly mature oocytes respond differently to mature oocytes when exposed to hypertonic insults, indicating alterations to membrane permeability and the cytoskeleton. The jelly layer in sea urchin oocytes spontaneously dissolves after a few minutes in seawater, and many synthetic processes may be initiated, precociously leading to the partial dissolution of cortical granules. In aged echinoderm oocytes, the elevation of the fertilization membrane is delayed by up to 60 seconds. Aging affects the oocyte's capacity to produce sufficient energy due to incompetent mitochondrial activity.

Metabolism and the Synthesis of Macromolecules

Fertilization induces many physiological changes in the oocyte, including permeability to small molecules, oxygen uptake, carbohydrate metabolism and synthesis of DNA, RNA and proteins (see Figure 5.15). Warburg in 1908 showed that the oxygen consumption of sea urchin oocytes increased six-fold following activation; however, this is not found at oocyte activation in most other animals. In sea urchins, this respiratory burst is preceded by an increase in the level of co-enzyme NADPH brought about by the phosphorylation of NAD. The low level of respiration in unfertilized sea urchin oocytes seems to be due to the low availability of the substrate glucose-6-phosphate. Similarly, in the sea urchin there is a 100-fold increase in protein synthesis after fertilization brought about by an acceleration in translation of stored maternal mRNA. In the sea urchin after fertilization, there is an increase in its permeability to amino acids and nucleosides.

During the final stages of activation there are many changes in macromolecules such as proteins and RNAs. Maternal proteins degrade, and new translation of maternal RNAs commence. Cytoplasmic polyadenylation involving elongation of the poly(A) tail after export of mRNAs to the cytoplasm occurs and is considered to be a regulatory mechanism for protein expression from specific mRNAs. A decrease in maternal mRNA in mouse embryos begins during the final stages of meiotic maturation. In fact, a decrease in translatable mRNAs and proteins continues until the time of the maternal to zygotic transition. It seems probable that microRNAs play a role in the regulatory process of selective degradation of mRNAs.

DNA and Protein Synthesis in Mammalian Oocytes

DNA synthesis in the mouse begins almost synchronously in the sperm and oocyte pronuclei about eight hours after fertilization, when the pronuclei have distinct nucleoli and is completed within the next eight hours; whereas in the human, DNA synthesis starts at 12 hours post-insemination. DNA synthesis can be induced in sperm by heterologous oocyte cytoplasm and in fact the timing of initiation of DNA synthesis seems to be determined by the oocyte species. Several new proteins are synthesized during the pronuclear stage probably using stored mRNA. Some are transiently synthesized appearing then disappearing. The rate of protein synthesis changes enormously during the late pronucleus stage

Mitochondria

The developing embryo must generate sufficient levels of energy to succeed in development. This energy formed by the oxidative phosphorylation of glucose leads to the production of adenosine triphosphate (ATP) and is produced mainly in the mitochondria. Mitochondria have an impermeable highly folded inner membrane and a highly permeable outer membrane. The human oocyte contains between 20,000 and 80,000 copies of mtDNA; however, this does not indicate either the number of functional mitochondria or their individual efficiency. In addition to their role in generating energy in the oocyte, mitochondria also act as a signalling mechanism that can control cell fate. They can modulate intracellular calcium dynamics and therefore affect gene expression while they also emit reactive oxygen species (ROS).

Meiosis Resumption and the Cell Cycle

The last phase of oocyte activation is the resumption and completion of meiosis, leading to polar body extrusion and cleavage, which seems to involve the decrease in activity of MPF and MAPK. The calcium increase at activation triggers changes in PKC, CAMK11, calcineurin. CAMK11 transduces the calcium signal, leading to the phosphorylation of Emi2/Erp1 and eventually to its proteolysis. This degradation may be followed by Cdk1/cyclin B inactivation. It is not clear how the calcium signal is transduced to the inactivation of MAPK. In some animals, the delayed increase in CSF may be related to the destruction of Mos and inactivation of MAPK. A large quantity of zinc, released at fertilization in an exocytotic event that correlates with the calcium waves, is also essential for cell cycle resumption possibly through a pathway involving Emi2 activity and Cdk1 and Cdc2 phosphorylation. The key event leading to the metaphase-anaphase transition is the interaction between Emi2/Erp 1, cyclin B and MPF.

Chapter 6

The Dynamics of Fertilization

A commonly held belief is that fertilization is a haphazard free for all with oocytes bombarded by hoards of fertilization-ready spermatozoa. This misconception arises from laboratory-derived images of oocytes being deprived of their investments and exposed to unnaturally high concentrations of spermatozoa. Under natural conditions, this is not the case – sperm to oocyte ratios at the site of fertilization are extremely low. Fertilization is not a first-order chemical reaction but a fine-tuned, gradual and controlled encounter of gametes, where each gamete, in order to progress, must receive a correct sequence of signals from its partner. In marine animals, whether or not fertilization occurs at all depends on environmental cues and chemotaxis, while in mammals, behavioural adaptations are required to ensure fertilization, such as the deposition of sperm in the female tract and the synchrony of mating. Regardless of whether spermatozoa are deposited in the vagina, e.g. humans, or directly into the uterus, e.g. horses and mice, only a minute fraction successfully migrate to the site of fertilization – and of those, only capacitated spermatozoa are competent. Capacitation is a transitory state and is thought to last one to four hours in the human with only 10 per cent being capacitated at any one time. To ensure the availability of fertile spermatozoa over several hours, when ovulation may occur, there is a continuous replacement of capacitated spermatozoa from the stored pool.

What Is Polyspermy?

In sea urchins and mammals, only one spermatozoon normally enters the oocyte. If more than one spermatozoon enters the oocyte, then the multiple nuclei interact together, leading to anomalous division – a term called pathological polyspermy. In other animals – e.g. ctenophores, insects, elasmobranchs, some amphibians, reptiles and birds – it is normal for several spermatozoa to enter the oocyte cytoplasm. This is called physiological polyspermy. However, here, only one male nucleus will interact with the female nucleus and the others degenerate. In a small minority of animals (about 0.1 per cent), the problem of polyspermy does not exist since fertilization is parthenogenetic and development proceeds without the participation of the spermatozoon.

Our knowledge on fertilization has been obtained mainly from using models, such as the sea urchin and the mammal, that lend themselves to experimentation in the laboratory. Gametes from these animals are easily harvested, and fertilization and early development are easy to follow in vitro. The textbook images of oocytes in the laboratory deprived of their extracellular coats with thousands of spermatozoa attached to their surface are familiar, and these led to the idea that a surface mechanism existed that allowed the entry of one spermatozoon while blocking the entry of supernumerary spermatozoa (Figure 6.1). In particular, two activation events, an electrical depolarization and the cortical reaction, were suggested to be mechanisms that had evolved to prevent the entry of multiple sperm. The term polyspermy prevention was born and quickly adopted into the scientific vocabulary, and this mystic

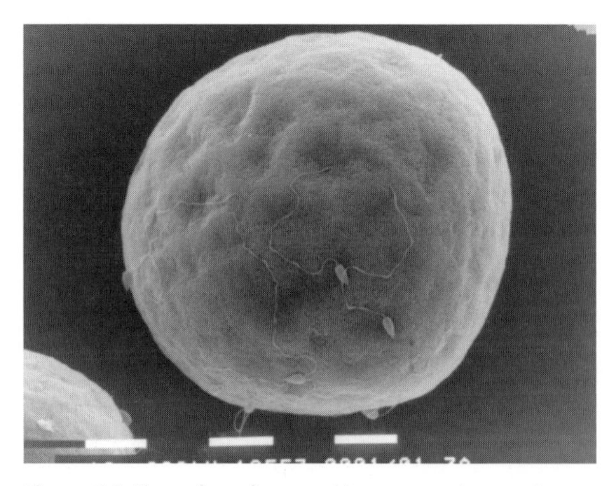

Figure 6.1 The surface of a sea urchin oocyte at the scanning electron microscope. Are all the attached spermatozoa capable of entering the oocyte, or is the fertilizing spermatozoon attached to a preferential site?

mechanism of sperm exclusion extrapolated to many other animal groups. However, under natural conditions, contrary to what many think, the number of spermatozoa at the site of fertilization is extremely low compared to the numbers generated. The sperm to oocyte ratio is regulated first by dilution in externally fertilizing species or the female reproductive tract in those with internal fertilization, followed by a bottleneck created by the oocytes extracellular coats, where only the fertilizing spermatozoon encounters and responds to the correct sequence of signals and enters the oocyte. Other spermatozoa are halted or slowed down by defective or sub-optimal signalling. Since under natural conditions, the final sperm to oocyte ratios approach unity, it would appear that selective pressures have favoured the achievement of monospermy rather than the evolution of polyspermy preventing mechanisms.

Laboratory Experiments and the Sea Urchin Model

Marine invertebrate oocytes have been the model of choice for centuries owing to their abundance and amenability to laboratory conditions and manipulations, and the experimental animal, par excellence, is of course the sea urchin, favoured by scientists of note such as Hertwig, Boveri and Herbst. Sea urchins are a member of the echinoderm family and are found globally. Animals are collected by divers often without taking into account their breeding patterns or sexual maturity. Gametes are obtained, either by injecting a concentrated salt solution into their body cavity or by manual dissection. The oocytes are surrounded by a jelly layer that is indiscriminately removed (by acid treatment), and these large cells are often attached electrostatically to slides to flatten them out, for both imaging and electrophysiological recording. Over 100 years ago, Delage showed that meiotic and cytoplasmic maturity in sea urchin oocytes is often asynchronous, leading to heterogeneous populations of oocytes. Although oocytes are effectively arrested at a specific point in the cell cycle as they wait for the activation signal from the spermatozoon to commence development, this metabolic block is not permanent, and oocytes in vitro will activate spontaneously – although at a slower rate than those activated by the spermatozoon. This is called aging and leads to the oocyte having an altered receptivity to spermatozoa, a condition that can be mimicked by treating oocytes with a variety of physical and chemical agents such as heat shock, nicotine and drugs that affect the actin cytoskeleton.

Images of sea urchin oocytes immersed in myriads of spermatozoa in a laboratory setting triggered the concept that fertilization, at least in these animals, was a haphazard 'free for all', analogous to a first-order chemical reaction and that the first spermatozoon that successfully interacted with the egg triggered membrane mechanisms that blocked the entry of other spermatozoa. Today, we are much more aware that scientific experiments need to be carried out under natural conditions and, in particular, to avoid manipulating oocytes and altering the characteristics of the cells before we start an experiment.

The Kinetic Experiments of Lord Rothschild

Lord Rothschild and colleagues in the 1940–1950s studied sperm–oocyte interaction in sea urchins at various concentrations and conditions, and came up with the idea that the fertilizing spermatozoon induced a fast yet partial change in the oocyte surface that preceded the cortical reaction and that reduced sperm receptivity by one-twentieth. Lord Rothschild made the assumption that a suspension of spermatozoa was analogous to an assembly of gas molecules, and he calculated the number of sperm–oocyte collisions at various sperm concentrations. By mixing oocytes and sperm at known densities for varying periods of time and by treating the fertilization reaction as a first-order chemical reaction, Rothschild found that the fraction of monospermic oocytes increased in time according to the relationship:

$$\mathrm{M}(t) = 1 - \mathrm{e}^{-\alpha t} \text{ below sperm densities of } 3 \times 10^6/ml.$$

The relationship is similar for polyspermic oocyte at densities between $7{\times}10^7$/ml and $3{\times}10^8$/ml, giving a rate of appearance of polyspermic oocytes as α^1. He argued that since α^1 (the re-fertilization rate) was found to be much less than α (the monospermic rate), a rapidly acting partial block reduced the probability of successful reactions after the first had occurred.

Although an exercise to be studied by all students of biology, the Rothschild hypothesis has fundamental flaws. First, fertilization is not a first-order chemical reaction and spermatozoa are not analogous to gas molecules. Second, it does not consider the possibility that there are a limited number of sperm entry sites on the oocyte surface. Lastly, it assumes that all spermatozoa are equal in their capacity to penetrate the cell. This is incorrect. Only competent spermatozoa that encounter and respond to the correct sequence of triggering events as they progress through the oocyte investments are successful.

Oocytes Are Programmed for Rapid Change

The oocyte and the spermatozoon are designed to trigger physiological changes in each other. In order to proceed through the layers of extracellular coats that surround the oocyte, the spermatozoon encounters a series of consecutive signals that induce changes in its physiology and are a prerequisite for further progression. If one observes sperm–oocyte interaction in echinoderms or mammals with their coats intact, it can be seen that many sperm are halted partially through their voyage. Once the spermatozoon has reached the oocyte plasma membrane, and probably following fusion of the two cells, this small cell (less than 500,000 times smaller than the oocyte) triggers the quiescent oocyte into metabolic activity that eventually leads to meiotic resumption and formation of the zygote.

The first signs of oocyte activation are changes in the ion permeability of the plasma membrane and the entry and subsequent release of intracellular calcium. While the former maybe localized to the site of sperm entry, the latter starts at the entry site and spreads in an autocatalytic wave to the antipode. Simultaneously, and probably in conjunction, there is a massive reorganization of the cell cortex involving cytoskeletal elements and a cascade of cell cycle kinases. The length and density of microvilli, location of cortical granules, and the priming of the intracellular Ca^{2+}-release mechanism are events in oocytes that are tightly controlled by the actin cortical cytoskeleton. In the starfish oocyte, the acrosome reaction can be seen to be induced at the outer layer of the jelly, and a long tubule then grows from the tip of the spermatozoon to perforate the jelly and the vitelline coat to fuse with the oocyte plasma membrane. The successful spermatozoon needs to encounter a series of consecutive signals that induce changes in its physiology and are a prerequisite for progression before triggering a localized sub-cortical polymerization of actin beneath the fertilization cone, which is then involved in the movement of the spermatozoon into the oocyte. This successful sperm-anchor site is regulated by the underlying actin cytoskeleton and, in any case, must be predetermined.

At activation, the oocyte changes its role from a quiescent cell programmed to interact with its male counterpart to a dynamic zygote that must be kick-started to progress through early development. The early embryo must be protected and isolated from its environment. Protection is afforded by the cortical reaction. Just below the surface of the sea urchin oocyte lie 20,000 granules, each measuring about one micron in diameter and containing a cocktail of enzymes and macromolecules (Figure 6.2). They

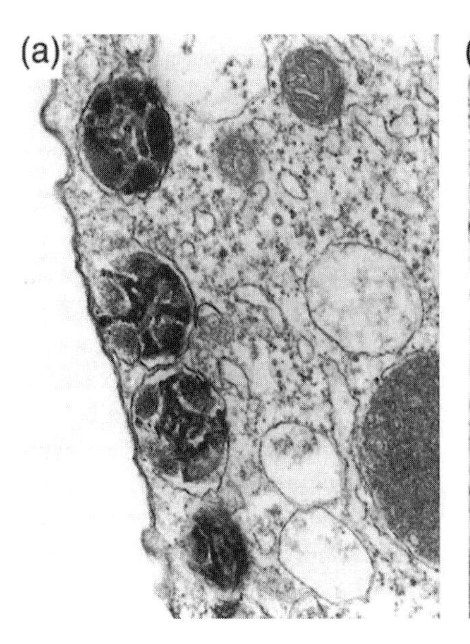

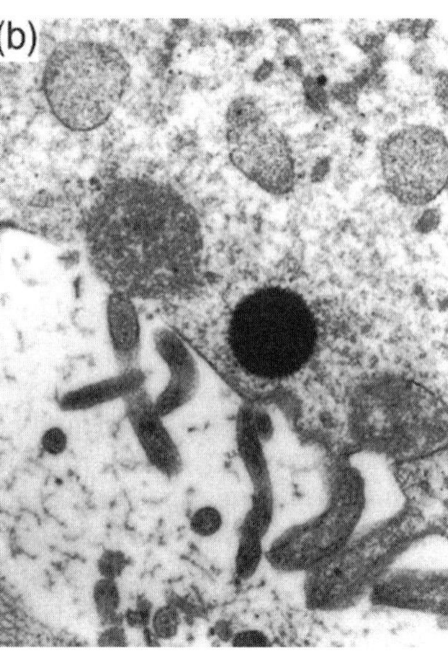

Figure 6.2 Transmission electron micrographs showing the cortical granules in an unfertilized sea urchin oocyte (left) and in a metaphase 2 human oocyte (right).

release their contents into the perivitelline space, following interaction with the fertilizing spermatozoon, by fusing to the oocyte plasma membrane. Fusion starts at the site of sperm entry and then traverses the oocyte in a wave to the antipode, taking about 30 seconds to complete. This leads to a net increase in the total surface area of the plasma membrane, which can be observed as a transient increase in length of microvilli. The cortical reaction has evolved to change the receptive outer investment of the oocyte into a hardened protective layer to protect the developing embryo in the early stages of embryogenesis and to provide a microclimate for the early division cycles and morphometric movements.

Activation events are propagative, predetermined and progress at a rate independent of sperm concentration. Thousands of biochemical and physiological pathways are triggered in a spatial and temporal pattern that does not lend itself to traditional molecular biological studies, such as depletion or over-expression of any particular component. These processes are not specific to gametes, but they are biochemical and physiological processes common to all cells, however with different kinetics.

Studying Sperm–Oocyte Dynamics Using Electrophysiology

The experiments of Lord Rothschild were the first indirect attempt to measure changes in the receptivity of the oocyte membrane to spermatozoa. However, it was not until the 1970s that electrophysiological recording gave a precise picture of these kinetics. Sea urchin oocytes are large and transparent and easy to impale with microelectrodes to record electrical changes across the oocyte plasma membrane. Many spermatozoa attach to the oocyte before one, the fertilizing spermatozoon, distinguishes itself by gyrating around its point of attachment. Approximately three seconds later, a small electrical step depolarization occurs, with no change in the morphology of the oocyte surface or sperm behaviour, until a further nine seconds later when a larger, bell-shaped depolarization starts (Figure 6.3).

The successful spermatozoon then stiffens and stops gyrating, and the cortical reaction is initiated at this point. A protuberance of the oocyte cortex (the fertilization cone) is formed and the sperm head slowly disappears into this cone, while the cortical

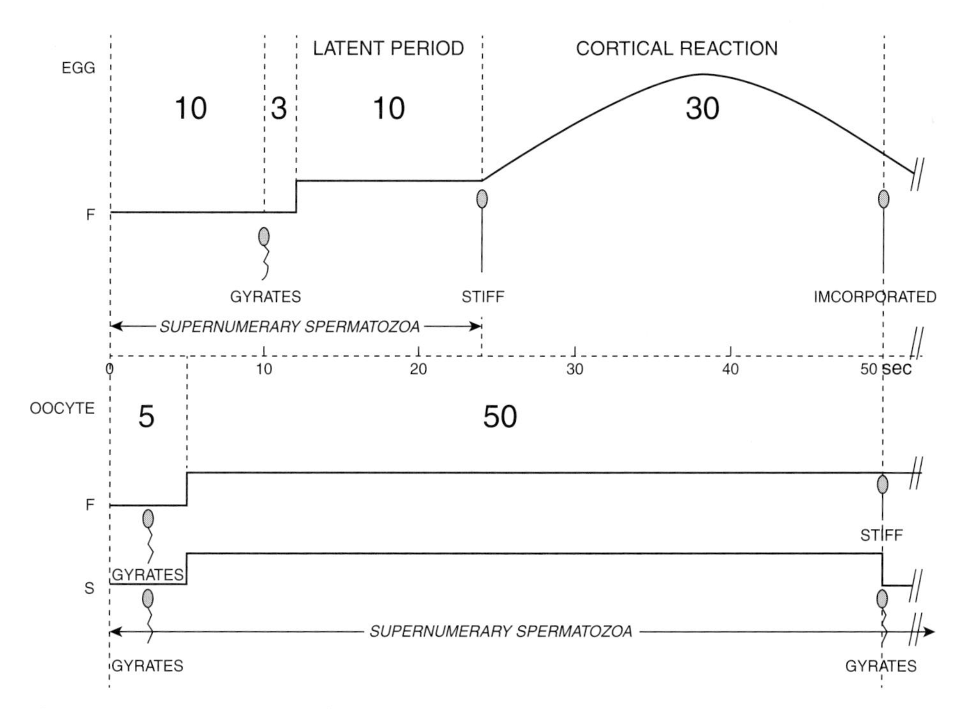

Figure 6.3 A schematic diagram of the voltage changes in a mature sea urchin oocyte (above) and a germinal vesicle sea urchin oocyte following insemination to show the behaviour of the fertilizing spermatozoa. Note that in both cases, supernumerary spermatozoa attach to the oocyte surface before and after the step event.

reaction spreads across the oocyte to the antipode. The cortical reaction leads to the elevation of the fertilization membrane and is completed during the repolarizing phase of the fertilization potential. Unsuccessful spermatozoa – that is, those that do not enter the oocyte – attach to the oocyte surface but do not generate either electrical step depolarizations or fertilization cones, and continue gyrating around their point of attachment for many seconds until their energy resources are depleted and fall limp to the oocyte surface. Thus progression of spermatozoa through the extracellular oocyte coats relies on them encountering and responding in sequence to oocyte-borne signals. It was suggested in the 1970s that the membrane depolarization, induced by the fertilizing spermatozoon, prevented the interaction of supernumerary spermatozoa, and this was called the fast electrical block to polyspermy. There is no evidence that depolarization prevents the interaction of supernumerary spermatozoa. Sperm entry is prevented at positive and negative potentials, in the voltage clamp configuration, however this is an artefact caused by the currents injected into the egg employed to hold the voltage constant in a non-physiological range. At permissive voltages – around -20 mV, where the current required to hold the voltage is minimal – only one spermatozoon normally enters the oocyte (Figure 6.4). Thus, irrespective of the oocyte voltage, the fertilizing spermatozoon is attached to a privileged interaction site that permits entry and distinguishes it from supernumerary spermatozoa.

The successful spermatozoon, which may also have predetermined characteristics, completes this sequential pathway, and the supernumerary spermatozoa fail at various stages, falling to the wayside. In either case, the oocyte and its topographical organization select the most fit spermatozoon. In physiological polyspermic species, such as the ctenophores, a parallel spermatozoon selection process by the oocyte occurs, but within the oocyte cytoplasm. In Beroe, for example, which has a very clear, large oocyte, several spermatozoa enter anywhere over the oocyte surface and remain immobile under the surface. The female pronucleus then starts to migrate around the oocyte surface, progressively encountering several sperm pronuclei until it decides to fuse with the successful sperm pronucleus. It is not clear how this selection process is regulated; however, here, as with pathological polyspermy, the fertilizing spermatozoon distinguishes itself from supernumerary spermatozoa.

Fertilization under Natural Conditions

Gametes vary enormously in form, function and numbers produced, and accordingly there are many ways in which sperm–oocyte interactions are regulated.

Animals with Low Sperm Numbers

Contrary to popular belief, not all animals produce large sperm to oocyte ratios. For example, in some insects and nematode worms, sperm utilization is very efficient. In *Drosophila*, there is a 1:1 ratio between the progeny produced and the number of sperm in the seminal receptacles. In the nematode worm *Caenorhabditis elegans*, every spermatozoon fertilizes an oocyte; however, not all oocytes are fertilized because in fact oocytes are produced in excess. In about 0.1 per cent of all animal species, reproduction is achieved without the participation of spermatozoa, and this phenomenon is called parthenogenesis. There are many variations in the precise strategy of parthenogenesis. In aphids, for example, parthenogenetic generations alternate with those generated by

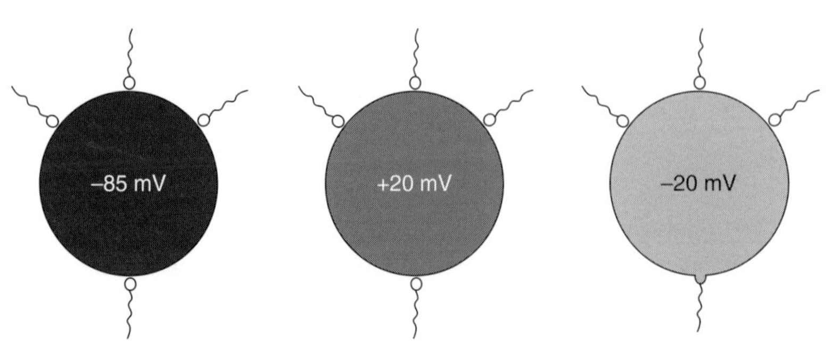

Figure 6.4 A schematic representation of voltage clamp experiments in the sea urchin oocyte. Holding the oocyte at extreme negative values (blue) or extreme positive values (red) inhibits interaction of all spermatozoa because of the non-physiological currents generated. Lowering the voltage to −20 mV (yellow) and consequently reducing the amount of current injected permits sperm fusion, but for only one – the pre-determined spermatozoon!

fertilization, and in bees an oocyte may be parthenogenetic or fertilized. The creation of female parthenogenetic offspring is widespread among insects such as some Phasmida, Diptera and Lepidoptera, while in the Crustacea, the best-known example of parthenogenesis is the brine shrimp *Artemia salina*. Nematodes, rotifers, snails and flatworms also include a few parthenogenetic species, whereas in the vertebrates, the most common form of parthenogenesis and female only species are found in the lizards.

Animals with High Sperm Numbers

In animals with high sperm numbers bottlenecks exist to reduce the number actually reaching the oocyte.

Reduction by Elimination

Mammals are classified as pathologically polyspermic, and the sperm to oocyte ratio at origin can be as high as 10^9:1. If we examine sperm size and number in relation to body mass in these animals, it appears that evolutionary responses have favoured sperm number rather than sperm size with increasing body size. Essentially, larger animals such as elephants produce more sperm per ejaculate (corrected for body mass) than a small mammal such as a mouse. Despite these large sperm numbers in the mammals, behavioural adaptations are required to ensure fertilization. Mating must be synchronized, and the sperm need to be deposited in the female tract. In humans, the spermatozoa are deposited in the vagina, while in mice they are deposited almost directly into the uterus. In both cases the vast majority of spermatozoa are rapidly eliminated from the tract. Only a minute fraction successfully migrate to the site of fertilization. The major barrier for sperm ascent in mice is the utero-tubal junction, with spermatozoa being progressively released from the lower part of the oviductal isthmus at ovulation. The first barrier to sperm ascent in humans is the highly folded mucus filled cervix, where sperm are retained and released over a period of several days. Contractile activity of the uterine wall aids sperm ascension to the lower isthmus, where they are sequestered until ovulation. Finally, migration from the isthmus to the ampullae appears to be due to sperm motility and contractile activity of the oviduct. In conclusion, in mammals, the number of sperm that reach the site of fertilization is regulated by the female tract. In addition, only a small proportion of spermatozoa at any time are capacitated and able to fertilize the oocyte. Spermatozoa only remain in a capacitated state for a few hours, after which they lose their fecundity. It should be noted that, as for capacitation, only a small percentage of spermatozoa are responsive to follicular fluid, and the chemotactic response is also transient. Consequently, there is a continuous turnover of fertile competent spermatozoa in a given population.

In the few in vivo studies where spermatozoa have been counted in situ, a few hundred spermatozoa were found in sheep ampullae, while only five were found in man. In rodents the sperm to oocyte ratios at the site of fertilization is usually one to one or below. Chemotaxis, although not universally accepted in the mammal may be essential to enhance gamete encounters considering this low number of sperm at the site of fertilization.

In birds, which are classified as physiologically polyspermic, hundreds of millions of sperm are inseminated, but only a few hundred reach the ovum; the vast majority are ejected by the female tract early after copulation. In the zebra finch and domestic fowl, the female tract regulates the number of sperm reaching the site of fertilization, and it has been suggested that although one or few spermatozoa are sufficient to activate the oocyte, the presence of several supernumerary spermatozoa in the cytoplasm of the oocyte is a prerequisite for embryogenesis.

In the Urodele *C. pyrrhogaster,* seven to eight spermatozoa normally enter the oocyte, and all de-condense to form sperm pronuclei with sperm asters radiating from their centrosomes. The female nucleus moves towards the centre of the animal hemisphere, and one of the sperm pronuclei moves there also. Why only one moves towards the female nucleus is unknown. The accessory sperm nuclei remain at the oocyte periphery, enter S phase and then degenerate before cleavage. The degradation of the sperm asters depends on factors in the oocyte cytoplasm, possibly MPF, which may be topographically distributed.

Dilution Bottleneck

In marine animals, whether or not fertilization occurs depends on environmental cues, behavioural adaptations and chemotaxis. Sperm concentration in the sea urchin testis is similar to that in mammals; however, many more oocytes are produced than in mammals. Consequently, the problem faced by these animals is making sure there are enough spermatozoa to satisfy all the oocytes spawned. In sea urchins, the sperm to oocyte ratio at source is 10^4:1.When dealing with sea urchin gametes in the laboratory, a sperm–oocyte ratio that leads to 100 per cent fertilization at time 0, with minimum rates of polyspermy is often selected (10^6/ml); however, data collected from natural spawnings show a much lower fertilization rate and great variability in the proportion of oocytes fertilized in nature. Fertilization success in nature depends on the spawning behaviour of the organisms, population size, current velocity, oocyte size, sperm swimming capacity and many other factors. The consensus from field studies is that fertilization success in free-spawning benthic organisms can be less than 1 per cent and is highly variable, ranging from 1 per cent to 95 per cent. Thus, in the environment, sperm–oocyte collisions are rare, sperm concentration maybe extremely low (below 10^4/ml), and the availability of sperm may affect female reproductive success. If, under natural conditions, sperm–oocyte collisions are indeed low, perhaps we should reconsider some of our ideas on the evolution of sexual dimorphism.

The Egg Coat Bottlenecks

In all animals, the oocyte is surrounded by complex extracellular layers or coats, which may be removed in vitro, often without inhibiting fertilization. This is often erroneously interpreted as showing the inutility of the extracellular coats and is purely a laboratory artefact. In nature, passage through these coats is a prerequisite for normal fertilization, oocyte activation and subsequent paternal nuclear de-condensation. Sea urchin oocytes are surrounded by a jelly layer composed of high molecular weight fucose-sulphate-rich glycoconjugates. At spawning, sea urchin sperm are exposed to the high pH of the seawater, which leads to the activation of a Na^+/H^+ exchanger and a dynein ATPase, causing an increase in motility. Factors released from the jelly layer, the sperm activating peptides (SAPs) further stimulate sperm motility and respiration. The majority of spermatozoa are unable to penetrate the jelly layer and remain immobilized at various depths within the jelly. Those that pass through must arrive in a physiological condition that both promotes binding and subsequent penetration of the vitelline membrane. Thus, in animals with external fertilization, sperm–oocyte ratios are extremely low and need to be enhanced by chemotactic mechanisms even to ensure that a minimum of oocytes are fertilized. The outer jelly layer attracts and traps spermatozoa; however, incompetent spermatozoa (whether it be due to the sperm or the oocyte) are prevented from progressing towards the oocyte.

We have also seen in mammals that few spermatozoa reach the oocyte under natural conditions; of those that do, some are not able to progress through the outer oocyte coating, the cumulus oophorus. A second extracellular coat, the zona pellucida, that impedes sperm progression even further is composed of several glycoproteins (ZP) that differ between species. The zona pellucida serves to modulate sperm binding and to protect the embryo during early development, but we know little about its topographical constitution and if indeed sperm entry is piloted to a specific site.

The most obvious morphological oocyte bottleneck is found in insects, squid and some teleosts. Here the oocytes lack cortical granules; the extracellular coat, the chorion, is impenetrable; and the spermatozoa lack an acrosome. In these species the spermatozoa enter the oocyte through a pre-formed entry site, the micropyle, where, in fish at least, a sperm attractant, a glycol-protein, has been identified as being responsible for guiding the spermatozoa to this area. In the anuran *Discoglossus pictus*, oocytes are highly polarized, with the animal pole marking the position of the meiotic plate and organelles distributed in a gradient towards the vegetal pole. Here, the spermatozoa may only enter through a restricted depression at the animal pole, called the dimple, where the fine structural organization is different to the rest of the oocyte surface. In tunicates, the spermatozoon enters the oocyte at a preferential site at the vegetal pole. Polarized sperm entry coincides with polarized oocyte activation events. The first event is the release of calcium from intracellular stores, which traverses the oocyte from the point of sperm entry to the antipode in a wave. For example, in jellyfish and the anuran *Xenopus laevis,* where the sperm enters the animal pole of the oocyte, the calcium wave starts at the animal pole and traverses the oocyte to the antipode. In ascidian and nemertean oocytes, the wave initiates at the vegetal pole, the site of sperm entry. In teleosts, where the spermatozoa are forced to enter at the animal pole through the micropyle, the calcium wave starts at the animal pole. Since sperm entry in some hydrozoans is confined to the animal pole and the spermatozoa lack acrosomes, perhaps the animal pole has a specialized structure. In mammals and echinoderms, the calcium wave is also initiated at the point of sperm entry; however, it is not known whether in these groups there are also preferential sperm entry sites.

Sperm–Oocyte Encounters Are Infrequent in Nature

In the laboratory, we are used to seeing images of oocytes inundated with hoards of spermatozoa, and this triggered the concept that fertilization was a haphazard 'free for all', analogous to a first-order chemical reaction and that the first spermatozoon that successfully interacted with the egg triggered membrane mechanisms that blocked the entry of other spermatozoa. This is a laboratory artefact. Activation events, triggered by the successful spermatozoon, occur at a rate independent of sperm concentration and serve to change the metabolism and morphology of the oocyte from a dormant cell to a dynamic zygote that needs to be protected from the environment throughout early development.

In nature, sperm to oocyte ratios are very low. In some species, they approach unity at source, while in others that produce high numbers of sperm, there is massive loss by elimination, dilution and fertilization. As spermatozoa progress through the extracellular layers, those encountering or responding to defective signalling fall by the wayside, until one, the successful spermatozoon, reaches its goal. If this is the case, selective pressures would favour the achievement of monospermy rather than the prevention of polyspermy.

Chapter 7

The Zygote and Early Embryo

In Chapter 5, we saw that the oocyte is transformed from a quiescent cell waiting to interact with the spermatozoon into a dynamic zygote that changes rapidly, both structurally and physiologically, and that this is induced by sperm activating factors. Whether these factors are soluble and are released into the oocyte cytoplasm, or are externally placed ligands, or both, has still to be clarified. In either case, the sperm-borne factors act on the oocyte while the body of the spermatozoon is still outside the oocyte cytoplasm. Two other paternal structures act from within the oocyte cytoplasm. The first is the centrosome, which provides the mechanism for cell division; the second of course is the paternal DNA. Studying sea urchins and roundworms at the turn of the century, Boveri, at the Stazione Zoologica in Naples, was the first to show that the male gamete provided the 'division centre' for the zygote. He further predicted that the centrosome is the cyclical reproducing organ of the cell.

During spermatogenesis, the sperm chromatin is tightly packed into a nuclear envelope that lacks pores with the histones, normally associated with DNA, being replaced by basic protamines, which are rich in arginine, serine and cysteine. This association with these highly charged basic amino acids coincides with the repression of transcriptional activity. After fertilization, once inside the oocyte cytoplasm, the male nucleus undergoes a reverse process of morphological and biochemical transformation. The protamines are lost, nuclear de-condensation is induced, and the male pronucleus is formed. During pronuclear maturation the paternal chromatin undergoes significant reorganization with active demethylation and histone acquisition.

We have seen in Chapter 4 that the interaction of spermatozoa with oocytes is species-specific and controlled at several levels of cellular recognition. In contrast, in the later stages of fertilization, sperm–oocyte fusion, sperm nucleus de-condensation, the cytoplasmic factors in the oocyte that promote the formation of the pronuclei and the timing of DNA synthesis are not species specific and can be induced in cross-species or even cross-phyla fertilization. At a certain stage in the early embryo, maternal stores of RNA and proteins will be depleted and need to be replaced by de novo transcripts generated by the new zygotic genome. This moment is termed ZGA (zygotic genome activation) and occurs mainly at the two-to-four-cell stage in the mouse and the four-to-eight-cell stage in the human.

Activation events not only serve to trigger the resumption of the cell cycle and meiosis, but also to induce the intracellular movement of maternal components in a process called cytoplasmic segregation. This leads to the formation of a polarized zygote that, following cleavage, which is the specialized cell division of the early embryo, will compartmentalize cytoplasmic domains into distinct blastomeres. Polarization is a fundamental event in all embryos and occurs sooner or later in early development. In some animals, such as *Drosophila*, embryonic axes formation is established before fertilization, while in ascidians and frogs, axes formation is established both before and after fertilization. In the nematode worm *Caenorhabditis*, axes formation is set up after fertilization.

Sperm Incorporation

The sea urchin is typical of most animals in that sperm–oocyte fusion occurs between the tip of the inner acrosomal membrane of the spermatozoon and the oocyte plasma membrane (Figure 7.1) by, as we have discussed in Chapter 5, a mechanism that still has yet to be determined. Microfilaments are involved in entry of the spermatozoon into the oocyte cytoplasm, and this can be demonstrated best in the starfish, where the acrosomal tubule is 25 μm long and can be seen under the light microscope (see Figure 4.8). After attachment, the fertilizing spermatozoon

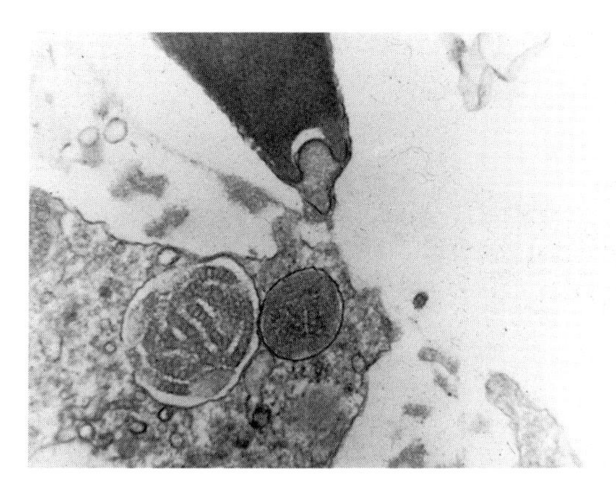

Figure 7.1 A transmission electron micrograph showing the extended acrosomal tubule of the sea urchin spermatozoon making contact with the oocyte plasma membrane. Note the cortical granule to the left of the spermatozoon is about 1 μm in diameter, and the contact area between the gametes is less.

continues flagellar movement for some 10–20 seconds then abruptly stops, and the tail becomes erect and perpendicular to the oocyte surface, which probably indicates membrane fusion between the two gametes. The fertilization membrane elevates around the tail, and, during the next 30 seconds, a protuberance of the oocyte surface, the fertilization cone, engulfs the spermatozoon, which moves into the oocyte cytoplasm at a rate of 5 μm/sec. Once in the cortex, the naked sperm nucleus moves laterally, rotates approximately 180 degrees, and during the next 10 minutes develops into the male pronucleus (Figure 7.2). The mitochondria and tail of the spermatozoon also enter the cytoplasm but later degenerate.

The process in small eutherian mammals is somewhat different and slower. Here the plasma membrane of the equatorial region of the acrosome-reacted spermatozoa appears to fuse with the oocyte plasma membrane. This mechanism of gamete fusion in mammals has often been the subject of controversy and needs further studies, preferably not using the electron microscope, which may introduce artefacts, to confirm this unusual fusion mechanism. If indeed correct, it is not clear how the equatorial segment of the spermatozoon becomes fusogenic, but since this is acquired after the acrosome reaction, it is possibly due to components released from the acrosome. F-actin has been observed in the equatorial region and post-acrosomal region of pig spermatozoa and may be involved in gamete fusion, as appears to be the case in starfish. Many components of the acrosome or sperm membrane have been implicated in sperm–egg fusion in mammals, including galactosyltransferase and various proteins including PH-30. The reader should refer to the Chapter 5, where sperm–oocyte fusion is discussed in the light of membrane lipid biology. In the small mammals, sperm do not fuse to the oocyte plasma membrane in the region over the meiotic plate, which is microvillus free; whereas in the human oocyte, which lacks a microvillus free area at the animal pole, it is not clear if there are preferential sites for sperm fusion (Figure 7.3). Also in mammals, one of the first indications of sperm–oocyte fusion is the abrupt cessation of flagellar movement: the tail stiffens and extends straight out from the sperm head. Sperm–oocyte fusion is quite advanced in small mammals after three minutes, the entire incorporation of the sperm head takes 15 minutes, and pronucleus formation takes about 60 minutes. The time sequence of fertilization events in the golden hamster is shown in Figure 7.4. In some mammals (for example the Chinese hamster) the tail is not incorporated, while in others it is incorporated by the progressive fusion of the oocyte and spermatozoon plasma membranes. After incorporation, the middle-piece mitochondria and axial filament of the tail appear to disintegrate. The spermatozoon plasma membrane, however, is integrated into the oocyte plasma membrane and may play a role in development. Sperm–oocyte fusion, in mammals at least, is not species specific, and spermatozoa from mice will fuse with oocytes from rats, guinea pigs and rabbits. The hamster oolemma will readily fuse with human and monkey spermatozoa. Gamete fusion is pH, temperature and Ca^{2+} dependent.

In the bovine, sperm incorporation and the conversion of sperm-derived components into the zygote have been explored after tagging sperm with a mitochondrion-specific vital dye. The zygotes were fixed at various times after fertilization for immunochemistry and ultrastructural studies. Results showed that complete incorporation of the sperm depends upon the integrity of oocyte microfilaments and is inhibited by cytochalasin B, which disrupts microfilaments. After sperm incorporation, the mitochondria were displaced from the sperm mid-piece, and the sperm centriole exposed to oocyte cytoplasm. The microtubule-based sperm aster was then formed, initiating union of male and female pronuclei. The disassembly of the sperm tail occurred as a series of precisely orchestrated events, involving the destruction and

(a)

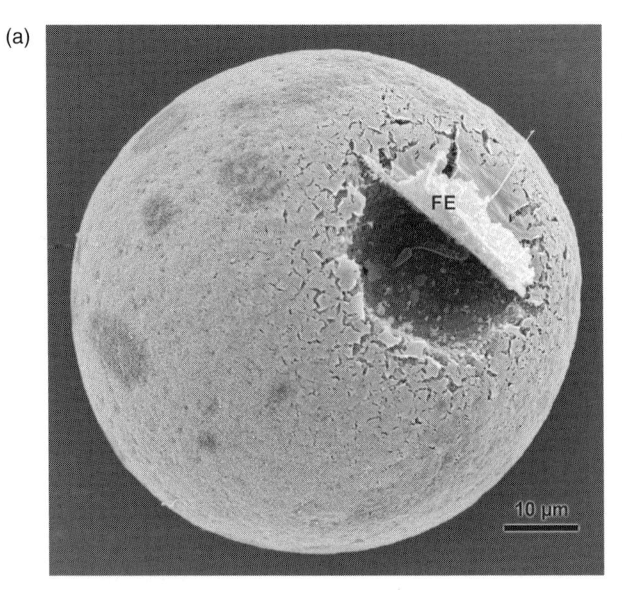

(b)

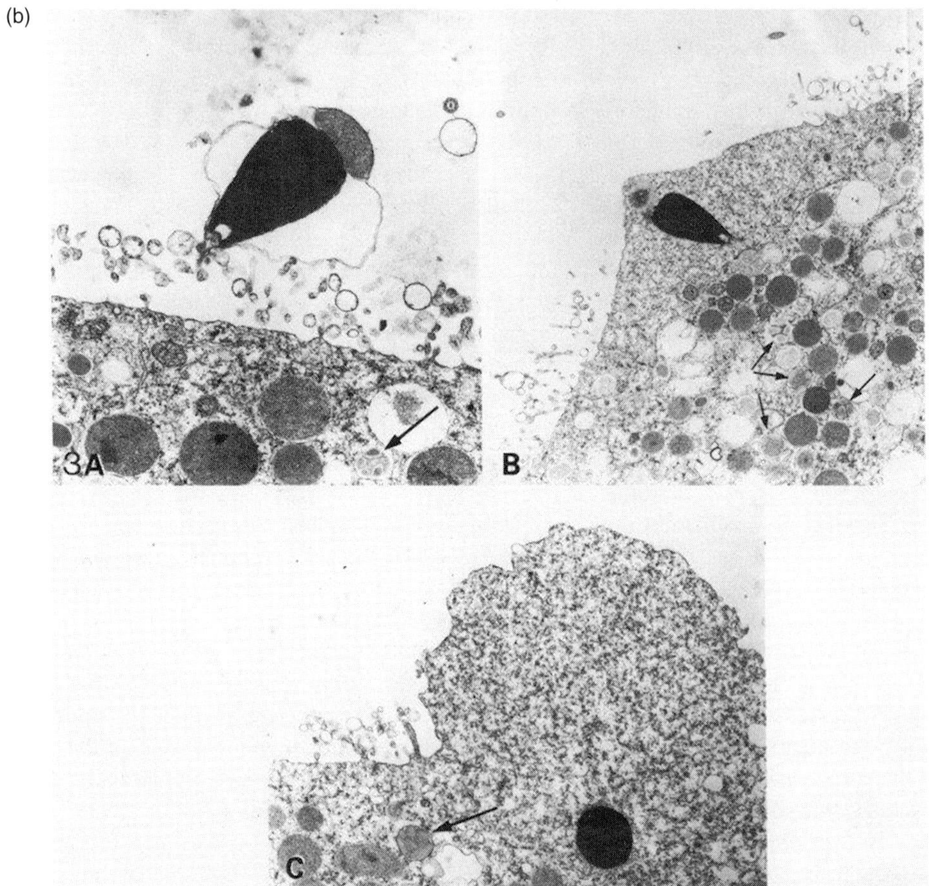

Figure 7.2 **(a)** A scanning electron micrograph showing the first moments of contact between the spermatozoon and the oocyte in the sea urchin. Note the sperm induces a concavity in the oocyte surface and the fertilization membrane starts to elevate before the spermatozoon is actually inside the oocyte (courtesy of Dr Luigia Santella, Stazione Zoologica). **(b)** A series of transmission electron micrographs in the sea urchin showing attachment of the spermatozoon to the oocyte surface, when it is incorporated in the fertilization cone and rotation of the sperm nucleus to 180 degrees to the oocyte surface.

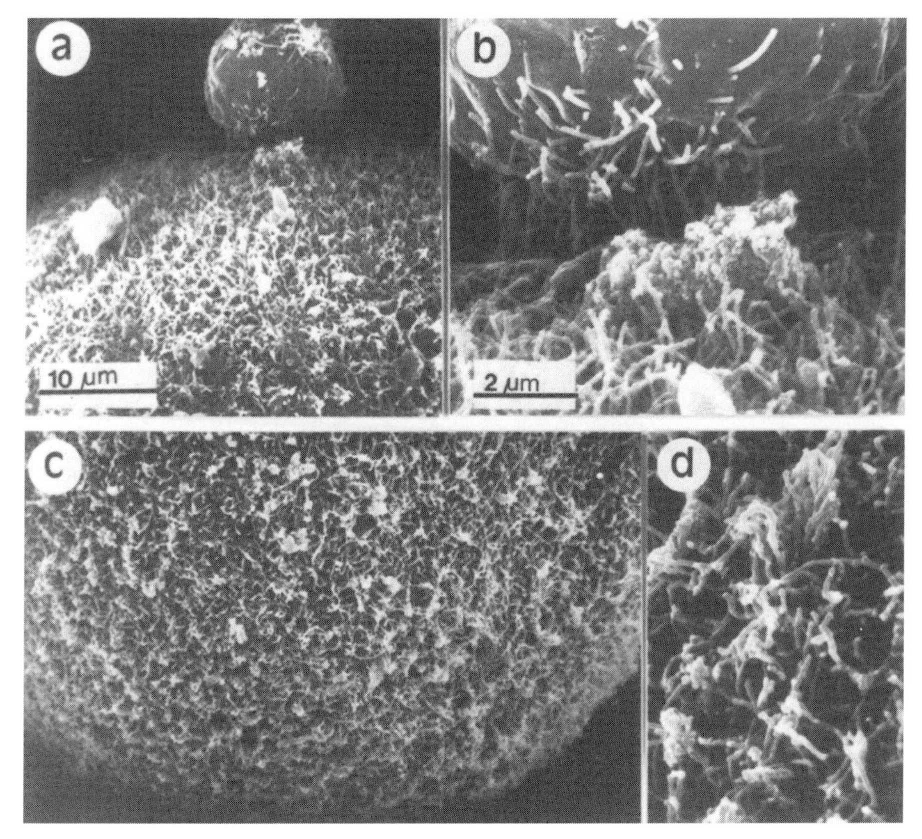

Figure 7.3 Scanning electron micrographs of the surface of the metaphase 2 human oocyte that show the homogeneous distribution of surface microvilli at the animal pole, where the polar body is emitted **(a)**, and the vegetal pole **(b)**. Rodent oocytes in contrast have a microvillus free area at the animal pole (from Santella et al. 1992).

transformation of particular sperm structures into zygotic and embryonic components. In mammals, some rRNAs, mRNAs, miRNAs and an endogenous reverse transcriptase are also carried by the spermatozoon into the oocyte cytoplasm.

De-condensation of the Sperm Nucleus

There are three basic processes involved in the development of the sperm pronucleus. Breakdown of the sperm nuclear envelope, de-condensation of the chromatin and development of the new pronuclear envelope. The sperm chromatin in some animals is not limited by a nuclear membrane; however, in most animals where it is present, after exposure to the oocyte cytoplasm, the inner and outer laminae of the sperm nuclear envelope fuse at multiple sites, forming vesicles that disperse in the oocyte cytoplasm. The nuclear lamina consists of a network of polymerized type-V intermediate filament proteins called lamins. Most of our information on sperm nuclear de-condensation has come from studies of the sea urchin, where the nuclear lamina, containing a 65 kDa B-type lamin, is first phosphorylated in a calcium dependent manner and disassembled by a cytosolic calcium-dependent protein kinase, PKC. Chromatin de-condensation then follows, which requires ATP hydrolysis, and transforms the conical sperm nucleus into a spherical structure. Sea urchin sperm DNA is packaged exclusively with histones not protamines. Two of these histones, Sp H1 and Sp H2B, only found in spermatozoa, are phosphorylated within minutes of fertilization which reduces their binding capacity with the DNA.

The first visible change in the sperm nucleus of mammals after penetration of the oocyte is the breakdown of the nuclear envelope, which starts at the equatorial region and proceeds anteriorly and posteriorly. The peri-nuclear material, called the theca, mingles with the ooplasm before de-condensation begins.

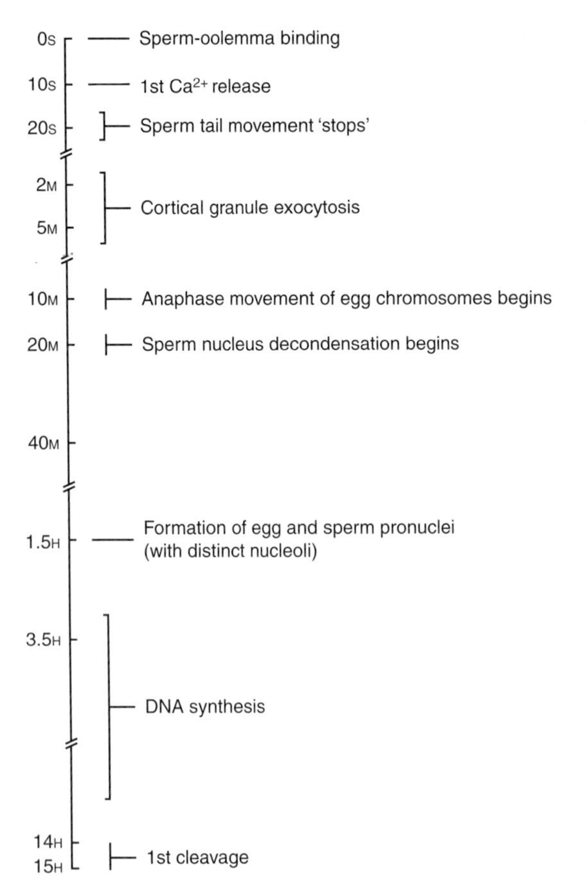

Figure 7.4 The time sequence of events at fertilization in the golden hamster (from Yanagamachi 1994).

In the golden hamster de-condensation can be seen to start 20 minutes after gamete fusion and is complete over the next 40 minutes. Sperm nuclear de-condensation is also not species specific, with human sperm nuclei being able to de-condense in hamster or indeed frog oocytes.

In mammals, the nuclear envelope, which lacks both a lamin lining and pores, is broken down by oocyte kinases, in particular PKC, exposing the contents to the ooplasm. De-condensation is initiated simultaneously with the reduction of disulphide bonds in the protamines, which relaxes the toroid complex allowing histones in the oocyte to associate with the paternal DNA. When histone deposition is complete the paternal chromatin expands fully into the male pronucleus. Dithiothreitol, an S-S reducing agent, will induce sperm nuclear de-condensation, suggesting that reduction of the S-S bonds in nuclear protamines and their replacement by histones is involved in sperm nucleus de-condensation. In eutherian mammals, the S-S bonds in protamines need to be reduced by reduced glutathione (GSH) in the ooplasm before they can be replaced by histones.

In the sperm nucleus of amphibians, protamines are replaced by histones within five minutes of it entering the oocyte cytoplasm. Nucleoplasmin (NPM), an acidic protein, the most abundant protein in the nuclear matrix, is released from the nucleus at GVBD and is thought to be responsible for the removal of protamines from the paternal DNA and the subsequent binding of histones. Nucleoplasmin–protamine binding is also not species specific, and DTT-treated human sperm nuclei will lose their protamines and de-condense within 30 minutes when exposed to toad nucleoplasmin. Mammals have three homologues of nucleoplasmin – NPM1, NPM2 and NPM3 – with the first and third being involved in sperm nuclear de-condensation. Nuclear reprogramming in mammals is also under the regulation of other proteins, such as NAP/SET proteins, the histone chaperone HIRA and the remodelling protein CHD1. The female DNA also requires de-condensation, although here the process is simpler in that the DNA remains associated with histones throughout. In the unfertilized oocyte of the mammal, the chromatin, blocked at the MII stage, remains in its condensed state until meiosis is complete and the second polar body is extruded.

Formation of the Pronuclei

In the sea urchin, the developing male pronucleus accumulates cleavage stage (CS) histone CS H1 from maternal stores, while the SP histone variants are degraded or lost. Two zones of the nuclear envelope remain attached to the chromatin following disassembly called lipophilic structures (LS), one at the base of the nucleus and one at the tip at the acrosomal fossa. Membrane vesicles in the oocyte cytoplasm initially bind to these areas, progressing towards the equator until they cover the entire surface of the chromatin. There are three populations of vesicle in the ooplasm that contribute to the formation of the male pronuclear envelope, MV1, MV2α and MV2β. Fusion of these vesicles is promoted by GTP hydrolysis and leads to the formation of a sealed envelope. The nucleus continues to swell, reaching 8–10 μm, by fusion of additional membrane vesicles and functional nuclear pores found in the cytosol, importing soluble lamins. This last process of swelling is

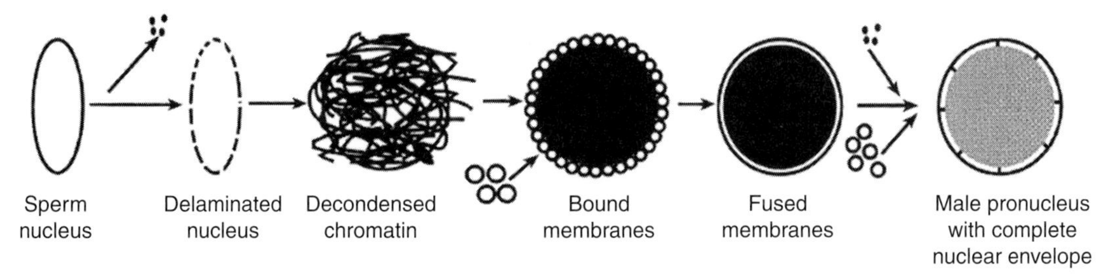

Figure 7.5 Formation of the male sea urchin pronucleus in vitro: (1) Sperm nuclear lamina is solubilized, mediated by PKC. (2) Chromatin decondenses inside the delaminated nucleus. (3) Membrane vesicles bind to chromatin. (4) Chromatin-bound membranes fuse. (5) Male pronucleus swells, and nuclear lamina assemble with formation of functional nuclear pores (modified after Collas and Poccia 1998).

dependent on calcium, ATP and GTP. A diagram illustrating pronuclear formation in the sea urchin is shown in Figure 7.5. In the zebra fish, a maternal protein called brambleberry regulates the fusion of chromatin masses surrounded by nuclear envelope into a large single nucleus.

In Chapter 2, we saw that oocytes from different animals are blocked at different stages of meiosis (see Figure 2.6). Consequently, the sperm nucleus, newly introduced into the oocyte cytoplasm, will find itself in a different cytoplasmic environment depending on the species. In oocytes that are normally fertilized at the germinal vesicle stage, such as in some molluscs, germinal vesicle breakdown occurs about 12 minutes after activation. The sperm nucleus de-condenses and then swells in conjunction with de-condensation of the female pronucleus. Insect oocytes are blocked at MI, and the sperm forms an aster when the oocyte reaches MII, but the paternal chromatin does not de-condense until the oocyte has completed both meiotic divisions. Vertebrate oocytes are fertilized at MII. In mammals, the first step of nuclear transformation is the reduction of the disulphide bonds in nuclear protamines. Dissolution of the sperm nuclear envelope and chromatin de-condensation occurs while the oocyte transits from metaphase 2 to telophase 2: during telophase 2, the sperm chromatin de-condenses as the female pronucleus develops. Sperm protamines are replaced by histones and the male and female pronuclear envelopes develop synchronously. Sea urchin oocytes are fertilized after the completion of meiosis, when the female pronucleus is already formed. Thus, in contrast to the other species, male pronuclear formation occurs in the interphase cytoplasm. Within two minutes of entry, the sperm nuclear envelope is disassembled, and within 30 minutes the nucleus swells some 20-fold to form the pronucleus.

Once formed, male and female pronuclei migrate towards each other and move towards the centre of the oocyte; the sperm aster is involved in this movement. In sea urchin and mouse oocytes, actin microfilaments and microtubules are required for pronuclear apposition, while in the human, only microtubules are required. In sea urchins, the envelopes of the two pronuclei fuse to form a zygote nuclear envelope containing de-condensed male and female chromatin. In most other animals (and as originally described for *Ascaris* by E. B. Wilson), the chromosomes in each pronucleus condense and concomitantly the pronuclear envelopes break down without fusing together. The male and female chromosomes then intermix in the cytoplasm and form the metaphase of the first mitotic spindle (Figure 7.6).

While sperm nuclear de-condensation does not require oocyte activation events, transformation of the male nuclear material into a pronucleus does require activation of the oocyte cytoplasm. Factors in the cytoplasm called pronucleus formative material (PFM) control the development of both male and female pronuclei in mammals and seems to be limited in quantity with the result that the male and female pronuclei compete for this factor. In the case of polyspermy, all the nuclei compete for these factors with the result that all nuclei are smaller in size. PFM's are not species specific or indeed cell specific since the nuclei of some somatic cells can be transformed into pronucleus-like structures when injected into mature oocytes.

In the mouse and rat, the sperm pronucleus is larger than the female pronucleus, which isn't the case in the human and golden hamster, where they are the same size. The endoplasmic reticulum is the major source of the nuclear envelope. There are two nuclear envelope precursors: NEP-B that binds to the

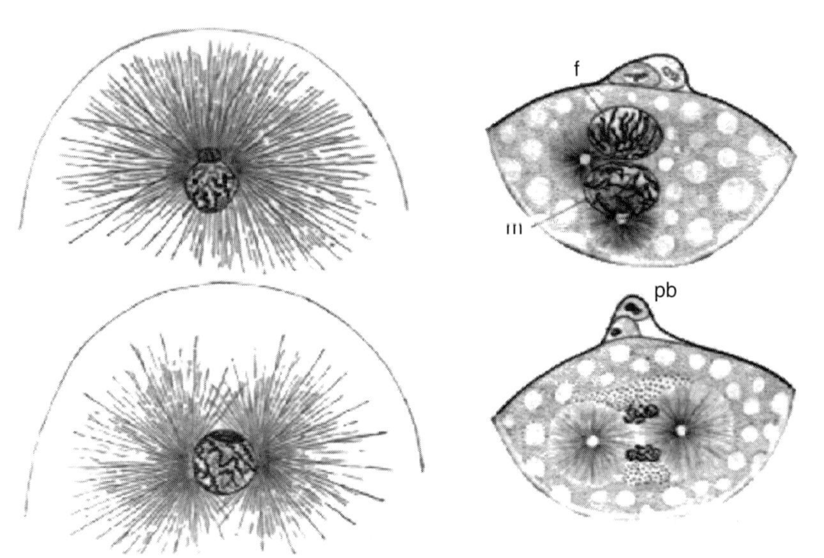

Figure 7.6 A drawing from Wilson (1900) showing (left) the fusion of male and female pronuclei to form the zygote nucleus as in sea urchins. In the majority of animals (and usually in mammals) the pronuclear membranes break down without fusing, allowing the chromosomes to interact in the cytoplasm (frames on the right).

chromatin and NEP-A that then binds to the chromatin-bound NEP-B.

The sperm pronucleus development factors are not species specific, human spermatozoa can develop into normal pronuclei in hamster oocytes and form a normal chromosome complement. The migration of the male and female pronuclei to the centre of the oocyte has been studied extensively, particularly in the sea urchin and the mouse. In the mouse, fluorescein conjugated probes for cytoskeletal elements show a thickened area of microfilaments below the cortex of the polar body region. In addition to the spindle microtubules, there are 16 cytoplasmic microtubule organizing centres or foci. Each centrosomal focus organizes an aster. Shortly before the nuclear envelopes disintegrate, the foci condense on the surface of the envelope and cleavage ensues.

Syngamy

Leopold Auerbach of Breslau, Germany (1828–1897) described two protoplasmic vacuoles in a newly fertilized egg, as well as a radiating figure between them. In 1876, Oscar Hertwig identified these vacuoles as the male and female pronuclei, and he observed their fusion. When the two nuclei merged together in syngamy, he described the figure: 'Es entsteht so vollständig das bild einer Sonne im Ei' ('It rises to completion like a sun within the egg'; see Elder and Dale 2011).

The migration of the sperm and oocyte pronuclei to the centre of the oocyte and their union is regulated by the behaviour of the cytoskeleton. Actins and fodrin, a spectrin-like protein, anchor the meiotic spindle to the oocyte cortex and draw the sperm nucleus deep into the oocyte. Microtubules are essential for the formation and migration of pronuclei. In the mouse, there are 16 cytoplasmic microtubule organizing centres all of maternal origin. Centrioles may also develop in rabbit blastomeres after parthenogenetic activation, also showing them to be of maternal origin. In the mouse, each centrosome focus organizes an aster following sperm incorporation, which then associate with the developing pronuclei. When the oocyte and sperm pronuclei are close to each other in the rabbit, the proximal surfaces become highly convoluted and the nuclear envelopes interlock and fuse at several places. In the human and mouse, the nuclear envelopes of the pronuclei breakdown separately before interlocking.

The Centrosome

The centrosome is composed of two centrioles placed at right angles to each other and surrounded by dense pericentriolar material. Each centriole is made up of nine triplets of microtubules arranged in a pinwheel array (Figure 7.7). During each cell division, in meiosis or mitosis, a new centrosome is assembled close to the original centrosome. Centriolar structure is highly

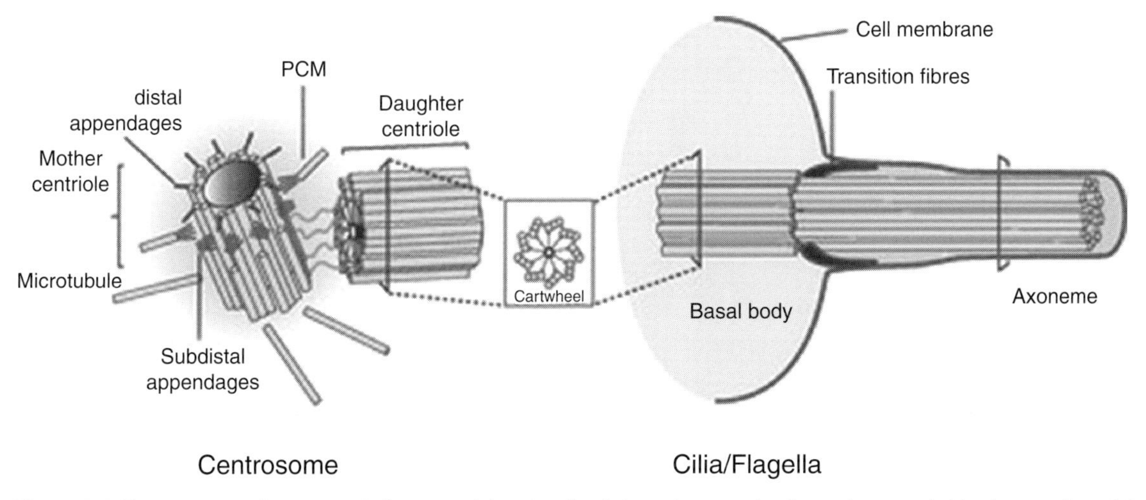

Figure 7.7 The centrosome is composed of two centrioles placed at right angles to each other and surrounded by dense pericentriolar material. Each centriole is made up of nine triplets of microtubules arranged in a pinwheel array (from Ross and Normark 2015).

conserved across the animal kingdom, except for the insects where the structure is extremely variable. Here the nine-fold symmetry is lost to a structure made up of thousands of microtubules (Figure 7.8) and may be related to the process of paternal genome elimination found in the male germline of insects.

In all animals, except the mouse, the centrosome is transmitted paternally to the zygote. The spermatozoon has a functional proximal centriole, which is close to the nucleus, and a degenerate distal centriole. Typically, a sperm-derived centriole induces centrosome assembly in the zygote. In almost all species, the female centrosome degenerates during oogenesis; whereas during spermatogenesis, the picture is more complicated with one, both, or none of the centrioles degenerating depending on the species. Where both centrioles are transmitted to the zygote, each induces the formation of a new centriole resulting in four centrioles which form the two centrosomes in the zygote. In other groups, including primates, one of the centrioles is highly degraded and therefore the competent centriole leads to the assembly of the four centrioles in the zygote. Finally, in rodents and some snails and insects, both male centrioles are lost, and the zygotic centrioles need to be all assembled de novo. Frog sperm centrosomes lack γ-tubulin, which is essential for the nucleation of microtubules. This is provided by the oocyte cytoplasm, and, therefore in frogs, the functional centrosome is a mosaic of paternal and maternal components. After the sperm enters the oocyte, an 'aster' of microtubules grows from the centriole, which directs the migration of the sperm pronucleus to the centre of the oocyte to make contact with the de-condensing maternal pronucleus, initiating its migration towards the forming male pronucleus (Figure 7.9). The zygotic centrosome then duplicates and splits apart during interphase, as microtubules extend from in between the eccentrically positioned, juxtaposed male and female pronuclei. After duplication, the centrioles migrate to opposite poles during mitotic prophase to set up the first mitotic spindle of the zygote. Although the centrioles are the main organelle associated with cell division, it is now thought that the pericentriolar material may be the principal microtubule-organizing centre.

Why centrosomes are degraded in the oocyte and are paternally transmitted remains a point of conjecture. Some argue that this is a mechanism to avoid spontaneous parthenogenesis, while others believe that both parents contributing a centrosome would complicate early embryogenesis. Excess centrosomes in the case of polyspermy are tolerated differently in different species. In some, this leads to total disorder, as in pathological polyspermy, while in physiologically polyspermic species, the excess centrosomes are degraded. Since the spermatozoon needs a centriole to form the axoneme, it is not really surprising that this cell contributes this structure to the zygote. It is important to note that most species are able to form centrosomes de novo from maternal proteins in the absence of a paternal copy.

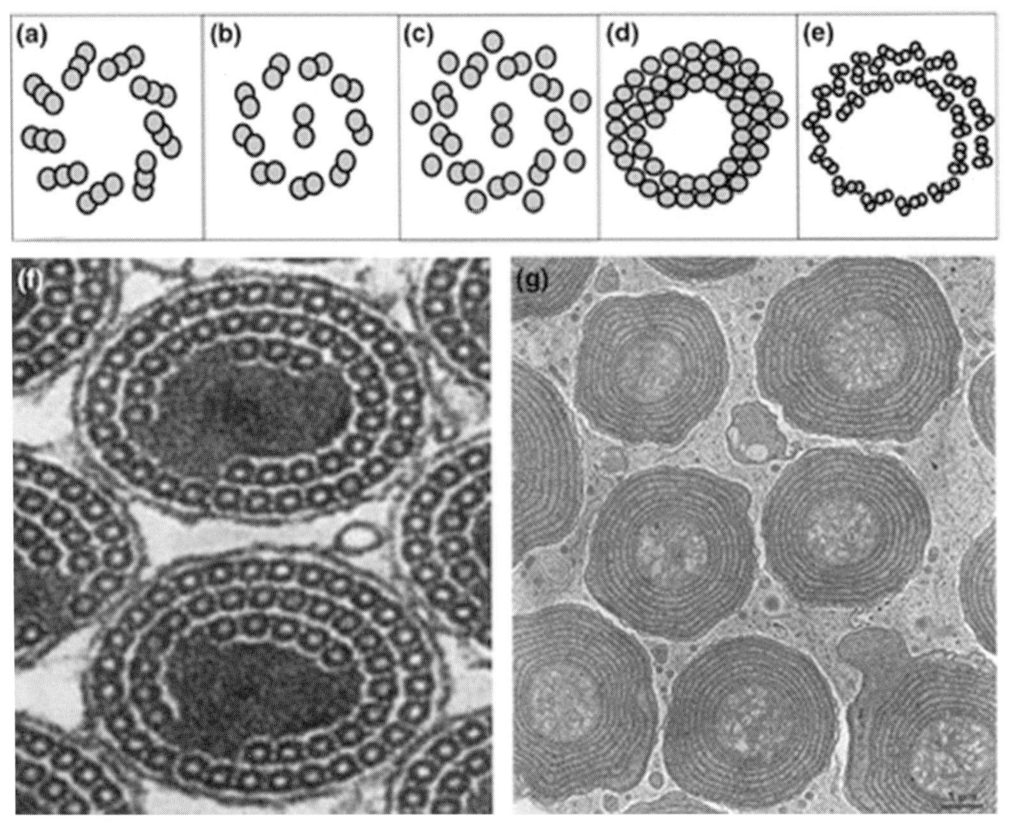

Figure 7.8 Centriolar structure is highly conserved across the animal kingdom, except for the insects where the structure is extremely variable. Here the nine-fold symmetry is lost to a structure made up of thousands of microtubules (from Ross and Normark 2015).

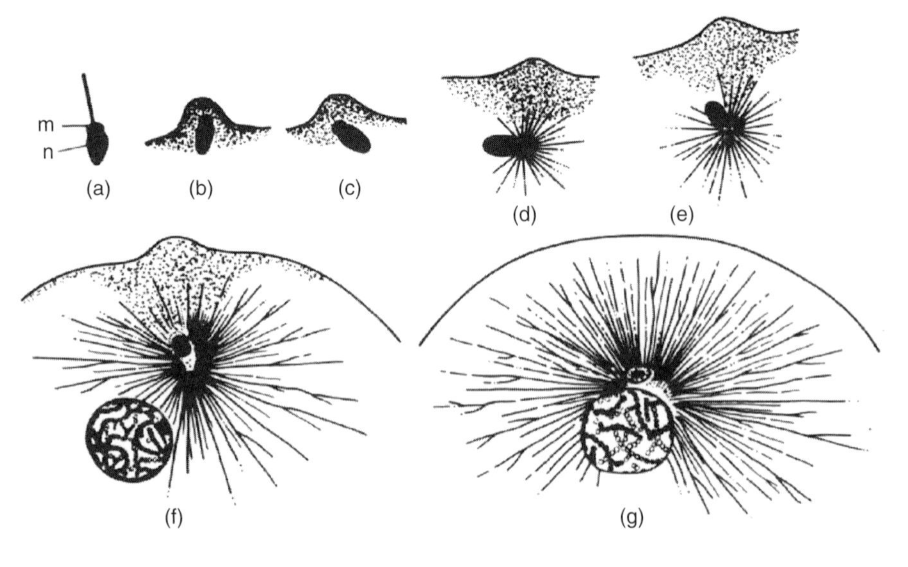

Figure 7.9 After the sperm enters the oocyte, an 'aster' of microtubules grows from the centriole, which directs the migration of the sperm pronucleus to the centre of the oocyte to make contact with the de-condensing maternal pronucleus, initiating its migration towards the forming male pronucleus (from Wilson 1900).

DNA and RNA and the Completion of Meiosis

In species in which the oocyte is blocked in meiosis to await the sperm, one of the first events visible through the light microscope after fertilization is the extrusion of a small cytoplasmic body called the polar body, which expels the unwanted contents of the female genome. The extrusion of the polar body occurs within five minutes of fertilization in ascidians, while it takes four to eight hours after sperm penetration in mammals. The sperm chromosomes are, at this point, a tiny speck in the oocyte cytoplasm and are distant from the oocyte chromosomes (Figure 7.10). The paternal DNA does not replicate until the female DNA completes meiosis, and the sperm DNA de-condenses. Many stages of meiosis are prone to error: the female genome may segregate erroneously during meiosis, leading to the retention or loss of single chromosomes within the cytoplasm, or segregation may not occur, leading to a triploid zygote. The male genome may fail to de-condense, blocking development. The cytoskeleton, responding to erroneous signals in the cytoplasm, may also fail to develop, again leading to the arrest of growth. In fact, examination of apparently unfertilized human eggs and abnormal zygotes with fluorescence microscopy reveals many of these anomalies in apparently unfertilized oocytes.

Both male and female pronuclei in mammalian zygotes undergo DNA synthesis generating the 4C amount of DNA required for mitosis. Prior to and during replication, the paternal DNA is globally demethylated, except for paternally imprinted genes, intracisternal A particle elements (IAPs), retrotransposons and centromeric regions all marks important for the development of the embryo. The maternal genome is protected from demethylation at this stage by the specific binding of the maternal factor STELLA or DPPA. In mouse zygotes, activation of the zygotic genome is first detected in mid S-phase and coincides with the massive degradation of maternal RNAs. Factors that control chromatin remodelling and DNA replication are also important for the regulation of zygotic gene expression.

The pronuclear envelopes breakdown and the chromosomes condense when MPF levels rise again. Cdk1 is responsible for the phosphorylation of the main A/C proteins, while poly(ADP-ribosylation) also appears to play a role in this process. The duplicated chromosomes now line up for the first mitotic division. Maternal mRNAs are recruited at activation, probably in response to the Ca^{2+} signal, and translated products confer transcriptional competence that underlies the activation of the zygotic genome. Other proteins that accumulate in mouse zygotes are associated with protein synthesis and degradation, in particular components of the ubiquitination pathway and the proteasome, which may assist with the

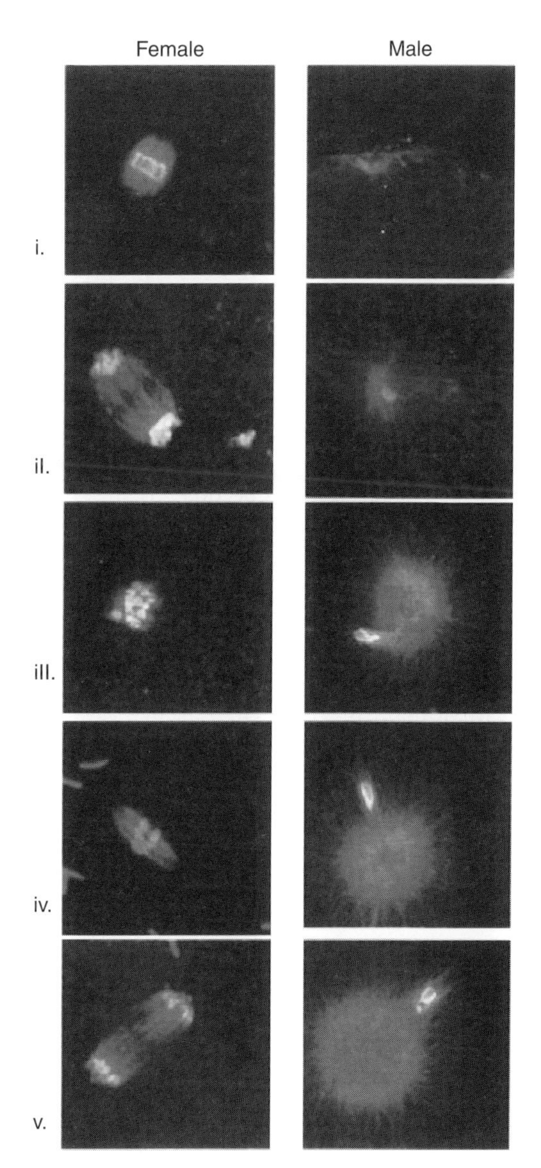

Figure 7.10 The formation and behaviour of the male and female centrosomal complexes after fertilization in the oocyte of the ascidian *Ciona intestinalis*. **i**. At sperm addition, **ii**. 5 minutes after fertilization, **iii**. 10 minutes after fertilization, **iv**. 15 minutes after fertilization and **v**. 25 minutes after fertilization (from Wilding et al. 2000).

degradation of maternal proteins that occurs at activation. The degradation of mRNAs is also essential for activation of the zygotic genome.

Cleavage and Mitosis

The zygote divides by mitosis into a number of smaller cells called blastomeres (Figure 7.11). This process of division, known as cleavage, is the opposite to the process of oogenesis in a sense: cleavage is a period of intense DNA replication and cell division in the absence of growth, whereas oogenesis is a period of growth without replication or division. Early cleavages are often synchronous, but sooner or later synchrony is lost. The blastomeres become organized in layers or groups, each group having a characteristic rate of cleavage. Although cleavage may be considered a mitotic process as found in adult somatic tissues, there is one important difference: in adult tissue, the daughter cells grow following each division and are not able to divide again until they have achieved the original size of the parent cell. The cells in a somatic population thus maintain an average size. During embryonic cleavage this is not the case: with each division, the resulting blastomeres are approximately half the size of the parent blastomere. As cleavage progresses, the embryo polarizes and differences arise between the blastomeres. Such differences may result from the unequal distribution of cytoplasmic components as already laid down in the oocyte during oogenesis, or from changes occurring in the blastomeres as a result of new embryonic gene transcription during development. Each blastomere nucleus will be subjected to a different cytoplasmic environment, which, in turn, may differentially influence the genome activity. As a result, after the onset of zygote gene activation and subsequent differentiation, eventually the blastomeres set off on their own particular programme of development to give rise to a particular cell line, for example nerve, muscle and so on. Although the first stages of development are pre-programmed and independent of external nutritional requirements, it must be noted that the embryo still has requirements for metabolites during this stage. Furthermore, simple mistakes and catastrophic events often arrest development, demonstrating that that the first stages of development are by no means certain to produce offspring.

The first cleavage plane in sea urchin embryos is (as in most animals) vertical, extending from the animal to the vegetal pole. The second division is also along the animal–vegetal axis (A–V axis), but at a right angle to the first. Thus at this stage there are four elongated blastomeres lying side to side. The third cleavage plane is equatorial (i.e. perpendicular to the first two planes), and the embryo is divided into

(a)

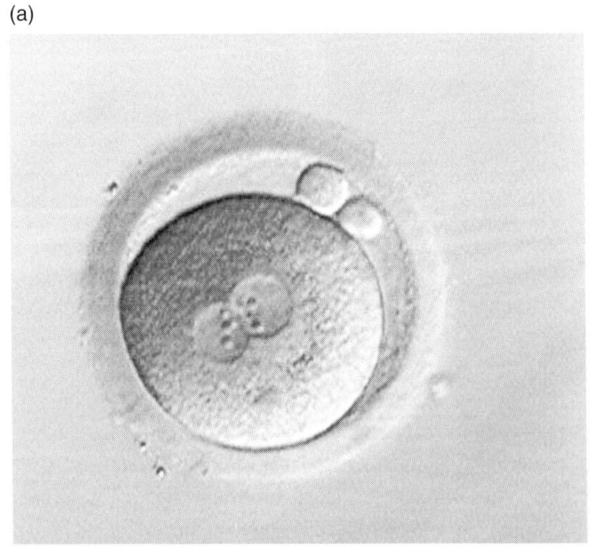

(b)

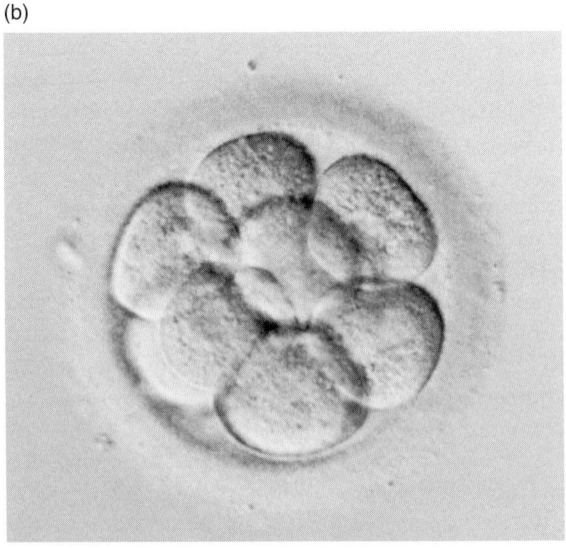

Figure 7.11 The zygote divides by mitotic divisions, called cleavage, into smaller cells the blastomeres. Cleavage is a period of cell division in the absence of growth, whereas oogenesis is a period of growth without replication or division. The images are of a human zygote, and an eight-cell stage embryo. (A black-and-white version of this figure will appear in some formats. For the colour version, please refer to the plate section.)

a tier of four animal blastomeres and a tier of four vegetal blastomeres. From now on the animal blastomeres cleave at one rate, the vegetal blastomeres at a different rate, and the blastomeres start to differ in size. The four animal blastomeres divide meridionally, forming a ring of eight mesomeres, while the four vegetal blastomeres cleave horizontally, with the cleavage plane shifted towards the vegetal pole. This latter cleavage division gives rise to four large macromeres and four tiny micromeres (Figure 7.12). Division continues, with the blastomeres becoming smaller and smaller until the embryo assumes a spherical shape called the blastula, which consists of a single layer of cells surrounding a liquid filled cavity, the blastocoel. The segregation of the four micromeres is of particular interest: these cells are committed to form the skeleton of the pluteus larva, and they also play an organizing role in embryogenesis.

Spiral cleavage, encountered in molluscs, annelids and nemerteans, is similar to the above pattern; however, the mitotic spindles are positioned obliquely with respect to the axis and equator of the blastomeres, and, as a consequence, the daughter blastomeres do not lie directly one above the other. For example, consider the third cleavage in the mollusc Trochus. Each of the four blastomeres divides into a small micromere and a large macromere. Thus, there is an upper tier of four micromeres (animal pole) and a lower tier of four macromeres (vegetal pole). The quartet of micromeres is shifted clockwise with respect to the macromeres so that the micromeres lie at the junctions of the lower cells rather than directly over them. Cleavage progresses in this elegant fashion, forming spirals of blastomeres until late in development, when the pattern becomes modified into one of bilateral symmetry (Figure 7.13).

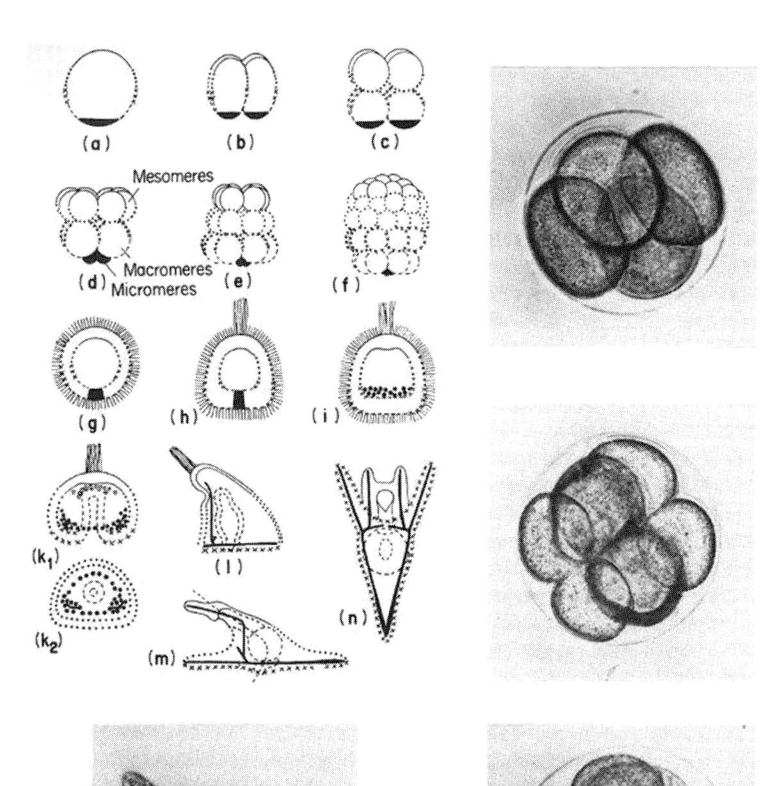

Figure 7.12 The first cleavage plane in the sea urchin embryo is vertical, running from the animal to the vegetal pole, while the second division is at right angles to the first, forming four elongated blastomeres lying side-to-side. The third cleavage plane is perpendicular to the first two planes, dividing the embryo into a tier of four animal blastomeres and a tier of four vegetal blastomeres. The four animal blastomeres divide meridionally, forming a ring of eight mesomeres, while the four vegetal blastomeres cleave horizontally with the cleavage plane shifted towards the vegetal pole. This latter cleavage division gives rise to four large macromeres and four tiny micromeres. These images are of 4-cell, 8-cell, 16-cell sea urchin embryos and the last is of the pluteus larva.

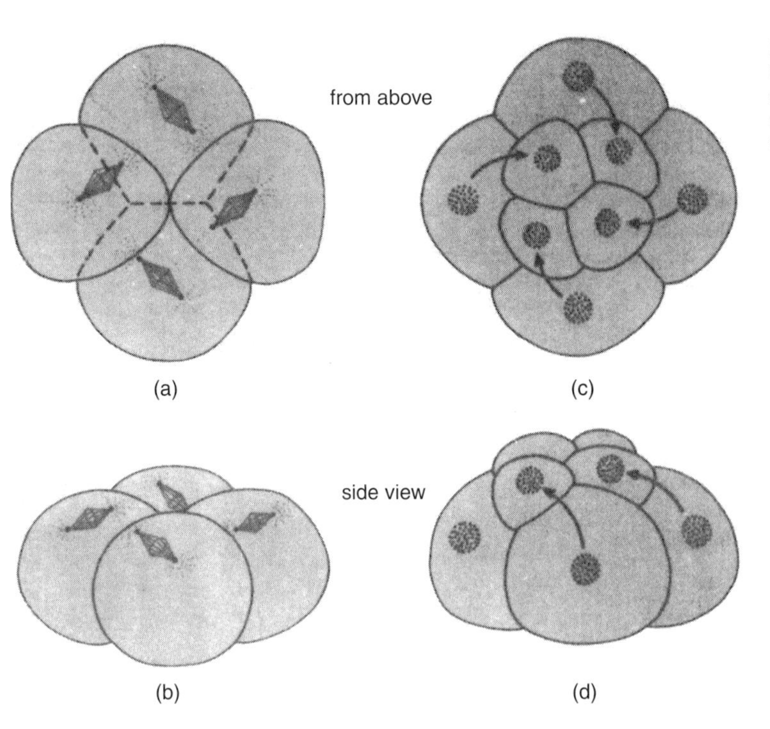

Figure 7.13 In molluscs and annelids following each cleavage the blastomeres are shifted clockwise so that they lie at the junctions of the lower cells not directly above them as in the sea urchin. This is termed spiral cleavage.

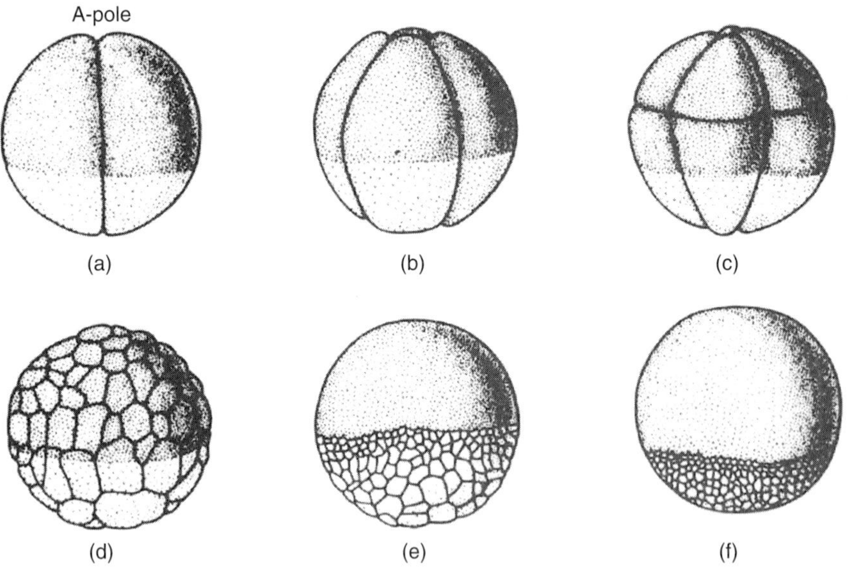

Figure 7.14 The pattern of cleavage is modified in yolk-rich oocytes, such as amphibian oocytes, where the yolk is distributed in a gradient with a maximum at the vegetal pole. Cleavage in these oocytes progresses faster at the animal pole than at the vegetal pole, and consequently the animal blastomeres increase in number and decrease in size at a faster rate than the vegetal blastomeres.

Large quantities of yolk tend to slow down the process of cleavage and consequently the pattern of cleavage is extensively modified in yolk-rich oocytes. In amphibian oocytes, the yolk is distributed in a gradient with a maximum at the vegetal pole. Cleavage in these oocytes progresses faster at the animal pole than at the vegetal pole and consequently the animal blastomeres increase in number and decrease in size at a faster rate than the vegetal blastomeres (Figure 7.14).

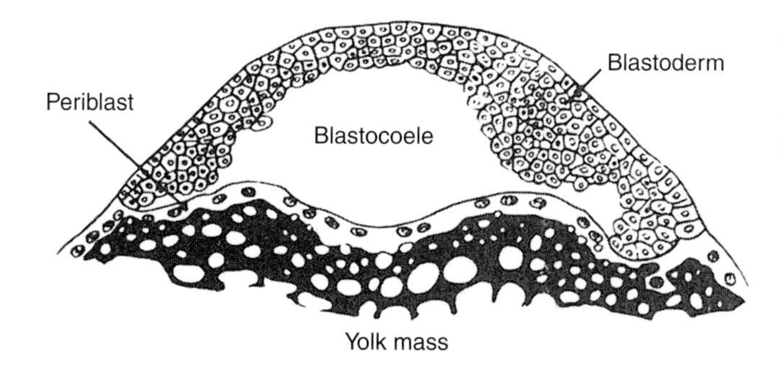

Figure 7.15 In fish, birds and reptiles, cleavages are incomplete or meroblastic, i.e. they pass through the restricted mass of cytoplasm located at the animal pole and not through the yolk. The embryo develops as a disc of cells perched on top of a yolk mass, the blastodisc.

In all the previous examples, cleavage is complete or holoblastic, dividing the entire oocyte into a number of small cells. In fish, birds and reptiles, cleavages are incomplete or meroblastic – i.e. they pass through the restricted mass of cytoplasm located at the animal pole and not through the yolk. The embryo develops as a disc of cells perched on top of a yolk mass, the blastodisc (Figure 7.15). Another type of incomplete cleavage is found in the oocytes of insects. Here, the zygote nucleus, lying in the centre of the oocyte, divides several times without the partitioning of the cytoplasm and the resulting nuclei migrate to the cell periphery. The peripheral cytoplasm segments around each nucleus and forms cells which remain in cytoplasmic contact with the central yolk body.

Cytoplasmic Segregation and the Formation of Cell Lines

In some animals, the cleavage pattern seems to be related to the constitution of the oocyte as laid down during oogenesis, in particular to the amount and distribution of yolk; in other cases, factors controlling the positioning of the mitotic spindles seem to be important. Whichever pattern is employed, cleavage causes a partitioning of cytoplasmic components, and this leads to differences in the developmental behaviour of blastomeres and the formation of different cell lines. Two classical examples, both marine invertebrate embryos, but quite different in their behaviour are described below.

E. G. Conklin in 1905 suggested that organ-forming substances were located in distinct regions of the ascidian zygote and called these domains 'plasms' (Figure 7.16). As cleavages proceeded, these plasms were segregated into specific blastomeres and gave rise to the various cell lineages. This basic concept was taken further by interpreting ablation and transplantation experiments in the 1950s by G. Reverberi and Giuseppina Ortolani in Palermo. In the late 1990s, Japanese scientists led by Norio Satoh and T. Sawada identified the cytoplasmic determinants for muscle (maternal mRNA, macho-1), the endoderm and the epiderm and showed they were already distributed in a gradient along the animal–vegetal axis. After fertilization these determinants are re-located and concentrated first along the dorsoventral axis and then along the anteroposterior axis.

In the unfertilized oocyte of *Phallusia mammillata*, a small actin-rich cap marks the animal pole above the spindle, and this is surrounded by a region lacking microfilaments. A sub-cortical domain, rich in mitochondria, 7–20 μm thick and lying just 1–3 μm from the surface is organized in a density gradient, with the highest concentration being located at the vegetal hemisphere. Conklin called this domain the myoplasm since the majority of this material segregates to the muscle cells of the tadpole. Bill Jeffery in the 1990s showed that ankyrin and myoplasmin C1, two cytoskeletal proteins, were predominant in the myoplasm as are a variety of mRNAs. A sub-cortical monolayer of ER tubes, which is attached to the plasma membrane binds the maternal postplasmic/PEM RNAs (called after their location at the posterior end mark in the posterior region of the zygote), PEM-1, macho-1, Hr-POPK-1 and PEM-3 and specific proteins. In many ascidians, vesicles and pigmented granules are also unevenly distributed along the A–V axis, which imparts a colour gradient easily visible under the light microscope.

The sub-cortical localization of mitochondria takes place during meiotic maturation over a period of two to three hours and is microfilament- but not microtubule-dependent, as is the movement of the

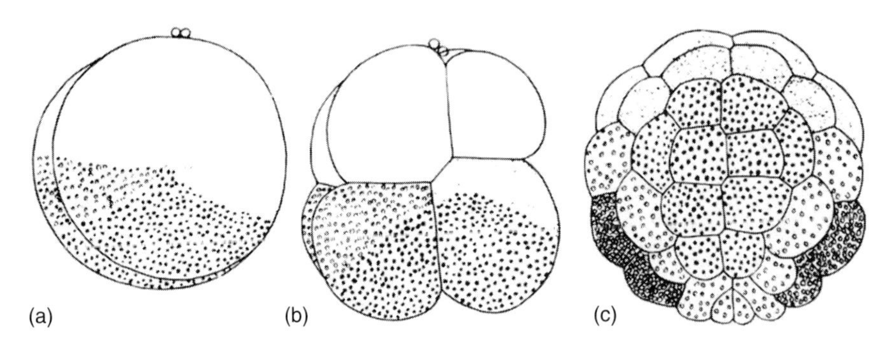

Figure 7.16 The segregation of plasms in the ascidian oocyte and embryo. The polar body marks the position of the animal pole. Progressive cell divisions compartmentalize the various preformed cytoplasmic domains (from Monroy and Moscona 1979).

centrally located meiotic apparatus that migrates to the opposite hemisphere of the oocyte. At about the same time some of the postplasmic/PEM RNAs also polarize along the A–V axis. At fertilization, there are two major reorganizations of the oocyte cortical cytoplasm. Shortly after fertilization, the contraction wave (see Chapter 5) driven by an actomyosin mechanism, starts at the site of sperm entry and spreads to the opposite hemisphere forming a bulge called the vegetal contraction pole. This contraction causes the myoplasm, the cortically located mRNAs, yolk and ER to concentrate at the vegetal hemisphere. The sperm nucleus and centrosome are also located at the vegetal pole. A second cytoplasmic reorganization occurs between the completion of meiosis and first cleavage, when the ER domain with associated RNAs and the myoplasm reposition at the posterior end of the zygote, which itself depends on the position of the sperm centrosome. This domain is now partitioned equally between the first two blastomeres and then asymmetrically over the next two cleavages, giving rise to a cortical structure called the centrosome attracting body (CAB). The CAB regulates the patterning of the posterior region of the embryo.

In some animals, such as *Drosophila*, embryonic axes formation is established before fertilization. In the ascidians, axes formation is established both before and after fertilization, as is the case of frogs, whereas in the nematode worm *Caenorhabditis* axes, formation are set up after fertilization.

In the sea urchin embryo, the organization of the early embryo is quite different. After fertilization, segregation of cytoplasmic components is less marked, and although cleavage results in the compartmentalization of the cytoplasm, the blastomeres retain a certain degree of flexibility. Thus, although each blastomere in situ develops into a certain part of the embryo, it retains the capacity to differentiate into other tissue types. This capacity can be demonstrated by simple experimentation. If a blastomere is isolated from a two- or four-cell embryo, it can reorganize itself and give rise to a whole larva. This phenomenon is called regulation. Amphibian and mammalian blastomeres may also be described as regulative. Blastomeres from mosaic embryos, such as the ascidian, do not have this capacity. For example, if the blastomeres of a two-cell ascidian embryo are physically separated, each blastomere gives rise to what it would have produced if left in situ, i.e. a half larva. There is a limit to the regulative capacity of sea urchin blastomeres, and a gradient of factors essential for normal development lies along the A–V axis of the unfertilized sea urchin oocyte. Blastomeres are only capable of regulation if they possess the vegetal factors. The third cleavage is equatorial, and if the animal blastomeres are separated from the vegetal blastomeres, only the latter cells are capable of forming a whole larva. The elegant implantation experiments of Horstadius in the 1920s and 1930s demonstrated that these vegetal factors are in fact distributed in a gradient. Essentially, in the 64-cell stage embryo, there are three tiers of vegetal blastomeres: those nearest the animal blastomeres are nominated veg 1, the next layer veg 2, and the micromeres form the distal layer. If an equatorial cut is made through the embryo, and the animal half isolated this will only give rise to a permanent blastula. However, if the micromeres from a second embryo are removed and implanted onto this isolated animal half, a normal embryo will develop (Figure 7.17). A similar though weaker effect is exerted by implanting the veg 2 blastomeres, but using the veg 1 blastomeres results in a defective larva.

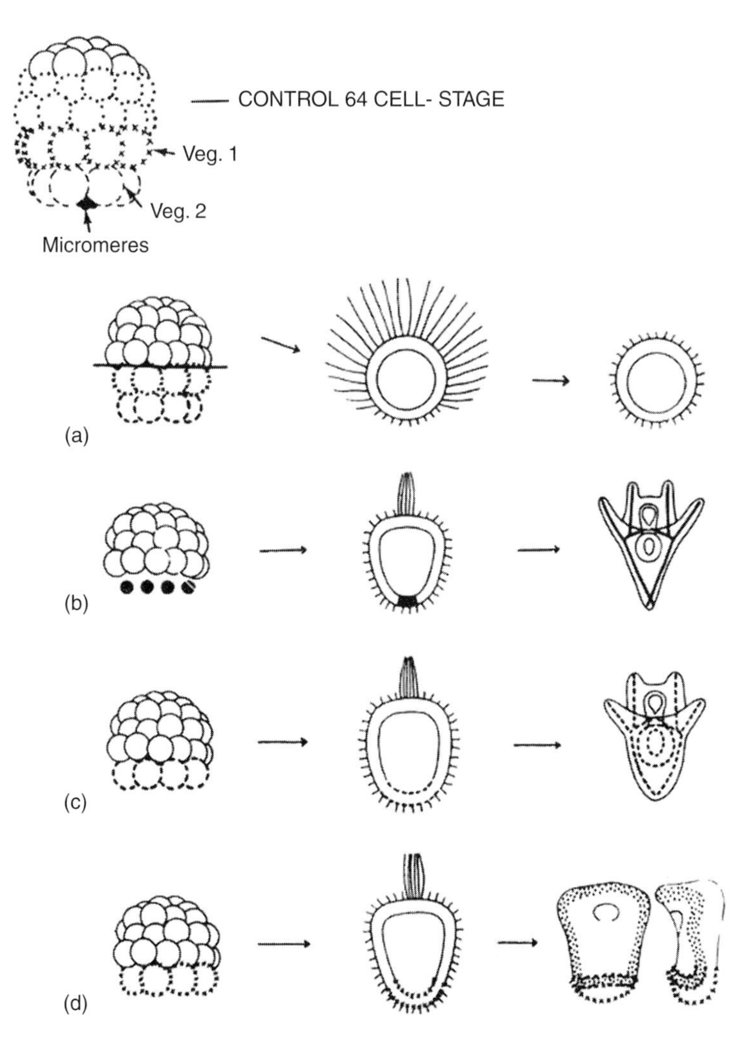

Figure 7.17 The implantation experiments of Horstadius 1939, demonstrated that vegetal factors in the sea urchin embryo are distributed along a gradient. If micromeres from an embryo are implanted onto an isolated animal half embryo, a normal embryo and pluteus will develop.

Zygotic Genome Activation (ZGA)

Successful development of a fertilized oocyte beyond the early cleavage stage requires *de novo* initiation and regulation of the new embryonic genome transcription. The oocyte cytoplasm forms a specialized environment for the newly formed zygotic nucleus. During the transition from maternal to zygotic gene activation, the embryo begins to synthesize its own RNA and protein rather than relying on that inherited from the mother. In the absence of appropriate activation and maintenance of embryonic gene expression, the embryo will simply fail to develop beyond early cleavage stages. This can be shown by inhibiting RNA production by transcriptional inhibitors and lead to the definition of the concept of the maternal-to-zygotic transition (MZT). MZT is not a single event but a gradual process that is not complete until all maternal stores of mRNA are depleted and the embryonic genome is fully transcriptional. MZT can be subdivided into two interrelated processes: first a subset of maternal mRNAs and proteins is eliminated, then zygotic transcription is initiated.

ZGA takes the embryo from a state of little transcription to one where thousands of genes are transcribed, which is very different to other developmental transitions where a cell's global transcription is relatively stable. Since cleavage is a mitotic event without cell growth, due to the large amounts of stored information in the oocyte, there is an enormous change in the ratio of the cytoplasmic volume to the nuclear volume during early development. In anamniote vertebrates, such as fish and amphibians, early embryonic divisions do not require transcription and rely on maternally supplied RNAs and proteins. Here, the cell divisions lack G_1 and G_2 gap phases, alternate between S and M phases and are rapid, occurring every 30 minutes, for example, in *Xenopus*. In zebrafish, these rapid, maternally driven cell cycles continue until the tenth division, while they persist up to the twelfth division in *Xenopus*. At this point, the S phase lengthens and gap phases appear, and this change in the cell cycle coincides with a significant increase in zygotic transcription. ZGA, however, consists of different genes being activated at different times, many preceding the late change in the cell cycle. In *Xenopus*, the first zygotic RNA to be transcribed is a microRNA (miRNA) at the eight-cell stage, which regulates the degradation of maternal mRNA. Transcription increases until the eleventh division cycle, when hundreds of genes initiate expression. In zebrafish, 600 zygotic transcripts have been detected by the 512-cell stage.

In contrast to fish and frogs, cell division in the early embryo of mammals is much slower, occurring every 12–24 hours. In the mouse, the first embryonic transcripts are detected in the male pronucleus, while transcription at the two-cell stage is RNA polymerase II (RNAPII) dependent and is required for successive cell divisions. A second larger phase of transcription starts at the four-to-eight-cell stage transition that marks morphological changes, which will lead to the blastocyst formation. In the human, ZGA is prominent at the four-to-eight-cell stage, a little later than the mouse, when up to 2,500 RNAPII-mediated transcription of genes occurs that are essential for cell division. The first transcripts detected in the human embryo are two paternal genes, ZFY and SRY, found at the pronucleate and two-cell stage. In the sea urchin, ZGA starts at the one-cell stage and transcription-factor-encoding transcripts are expressed in distinct waves at the two-cell stage, the early blastula and the early gastrula. An interesting fact in the sea urchin is that although transcripts encoding signalling receptors are present in the maternal pool, those encoding their ligands are strictly zygotic, possibly because the latter are expressed in a spatially regulated manner to specify cell fate and position in the embryo.

There are at least three models to explain the timing of ZGA. In the first, an accumulation of maternally deposited activating transcription factors are thought to trigger MZT, while others believe the nuclear/cytoplasmic volume ratio is important. Alternative views believe the state of the chromatin, which is permissive for transcription is decisive. ZGA may well be orchestrated by all three mechanisms and probably others as yet not identified.

The effects of the nucleo/cytoplasmic ratio on the control of embryology is not new and has been studied for over a century. Oocytes are enormous cells owing to the vast quantities of stored maternal components. At each cleavage division, the cellular volume is decreased by half, which doubles the nuclear to cytoplasmic volume ratio. The hypothesis is based on the concept that a ZGA repressor in the cytoplasm is titrated by the increasing number of nuclei relative to the unchanging volume of the cytoplasm. The idea was supported by the observation that ZGA in *Xenopus* is not based on a specific number of cleavages, rounds of DNA replication or a 'clock mechanism'. The most likely nuclear component involved in this nuclear/cytoplasmic ratio hypothesis is DNA. In anamniotes, such as fish and frogs, the sharp transition from synchronous to asynchronous cell cycles termed the mid-blastula transition (MBT) appears to be controlled by the N/C ratio. One model explaining the way the N/C ratio might work implies that the increasing DNA content at each round of division titrates away maternal inhibitory factors. Titration of histones and/or DNA replication factors may well regulate MBT.

In the activator model, the maternal mRNA of a transcriptional activator is translated. The activator then accumulates and targets gene activation depending on the amount of time it takes to reach a critical concentration. In *Drosophila* embryos, a specific ZGA activator has been identified as the transcription factor Zelda, while in zebrafish the transcription factors Nanog, SoxB1 and Oct4 are required to initiate the major phase of ZGA. In mammals, general transcription factors such as TIF1α, Yap1 and DUX4 human homeobox transcription factors appear to be involved in ZGA.Finally, various chromatin and methylation states are also likely to influence transcriptional competence and ZGA.

The onset of ZGA is critical for embryo development, at least in the mammal. In fact, one of the hypotheses of the mechanism of the early developmental block of mammalian embryos suggests that the early depletion of maternal stores prior to ZGA is a major determinant of embryo arrest and this depends on the 'quality' of oocytes produced during maturation, i.e. maturation is a vital determinant of embryo viability.

Polarization and the Formation of the Blastocyst

Initial cell fate decisions and morphogenetic movements depend upon the radial polarization of early blastomeres. Radial polarity maybe generated as a result of cell–cell interactions via cell surface adhesion proteins such as E-cadherin, or by asymmetric trafficking of membrane components leading to distinct membrane areas. Radial polarization is often the first step in the formation of an epithelial layer, with cells having distinct apical and basolateral surfaces and junctional complexes. Cell surface signals induce the asymmetric localization of polarity regulators, which include the apical PAR proteins (PAR-3, PAR-6) and PKC (aPKC), the apical Crumbs complex and the basolateral Scribble complex (Figure 7.18). Perhaps the most important polarity regulator is PAR-3, which is required for the localization of all the three groups of proteins. Polarity regulators promote the formation of tight junctions which contain the protein ZO-1 and Claudins. Finally, adherens junctions, which form between cells to promote adhesion, contain members of the cadherin family and catenin adaptors that couple the cadherin cytoplasmic tail to cortical F-actin and signalling proteins. The expression of radial polarity can occur in an embryo well before the formation of an epithelium. For example in the worm *C. elegans,* blastomeres radially polarize at the four-cell stage, in *Xenopus* at the two-cell stage and in mice at the eight-cell stage. Polarized blastomeres of the worm *C. elegans* do not form junctions with each other or develop into epithelial cells; however, polarity is required for cell movements in gastrulation.

In mouse eight-cell embryos, the blastomeres compact by increasing adhesion causing the cell surface to flatten. A distinct apical domain is formed containing microvilli and enriched in actin and actin binding proteins. During the next rounds of division, some cells divide asymmetrically, causing one daughter cell to occupy an external position and the other an internal location. Tight junctions then develop between the outer cells, marking radial polarization, and lead to the formation of the liquid-filled blastocoel cavity. Cell contact is necessary for radial polarization in the mouse. Before polarization, mouse blastomeres are totipotent, but after polarization, the inner and outer cell populations express different subsets of transcription factors and have different fates. The outer cells express Cdx2 and differentiate into the trophectoderm, while inner cells express Sox2, Nanog and Oct4 and become the inner cell mass. The ICM remains pluripotent and develops into the embryo and the extraembryonic endoderm.

It is not clear what external cues induce radial polarity in the mouse embryo; however, one candidate is the adhesion protein E-cadherin, which may transduce polarity information from the cell surface to the

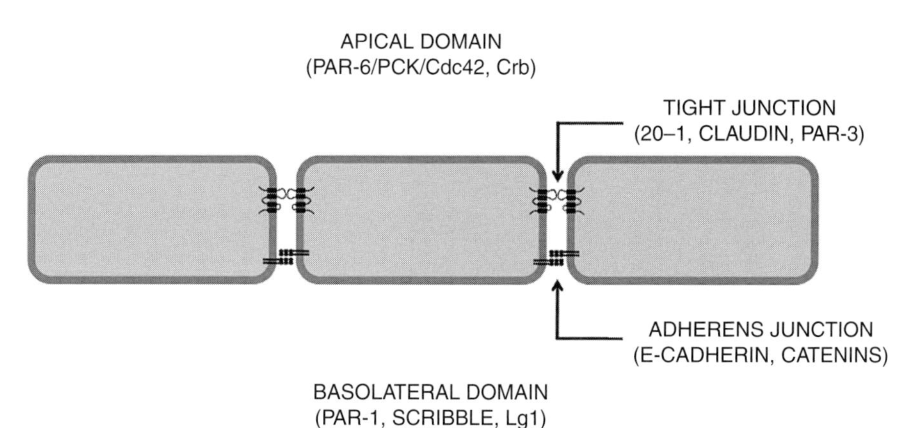

Figure 7.18 Radial polarization is often the first step in the formation of an epithelial layer with cells having distinct apical and basolateral surfaces and junctional complexes. Cell surface signals induce the asymmetric localization of polarity regulators, which include the apical PAR proteins (PAR-3, PAR-6) and PKC (aPKC), the apical Crumbs complex (Crb) and the basolateral Scribble complex.

adjacent cortex. In *C. elegans,* cortical polarity is broken by the PAC-1 mediated asymmetry in Rho GTPase, signalling between contacted and contact-free surfaces. In mice, the Hippo-signalling pathway may translate the state of polarity into cell fate. This signalling cascade responds to cell interactions to regulate the nuclear localization of the transcriptional co activator Yap. On activation of Hippo signalling, Yap is phosphorylated and localizes to the cytoplasm, when the pathway is inactive, Yap localizes to the nucleus. In the early mouse embryo, the Hippo pathway is off in outside cells and Yap is nuclear, while it is on in inside cells where the Yap is cytoplasmic. The cortical protein Amot is found asymmetrically and is required to activate the Hippo pathway. Amot can interact with actin, E-cadherin and Yap and may integrate cell contact information with Hippo-signalling activity.

There are two types of intercellular junction in early embryos; tight junctions and desmosomes, which anchor the cells together and form a permeability seal between cells and gap junctions. This allows the flow of electrical current and the transfer of small molecules, including metabolites and second messengers (cAMP), between blastomeres.

The surface of unfertilized human oocytes is organized into dense evenly distributed long microvilli which decrease in length after fertilization and even more so at the two- to eight-cell stages (Figure 7.19). There is no obvious change until around day four, when the microvilli are distributed in a polarized fashion over the free surface of the compacted blastomeres. In human embryos, tight junctions begin to appear at the six- to ten-cell stage (Figure 7.20), at the onset of compaction, while gap junctions are not well developed until the early blastocyst stage, when intercellular communication is clearly seen between ICM cells (Figure 7.21). ICM cells preferentially communicate with each other and not with trophectoderm cells via gap junctions, whereas trophectoderm cells communicate with each other and not with ICM cells. Between the 16- and 32-cell stages, activation of a Na^+/K^+ membrane pump results in the energy-dependent active pumping of sodium into the central area of the embryo, followed by the osmotically driven movement of water to form a fluid-filled cavity, the

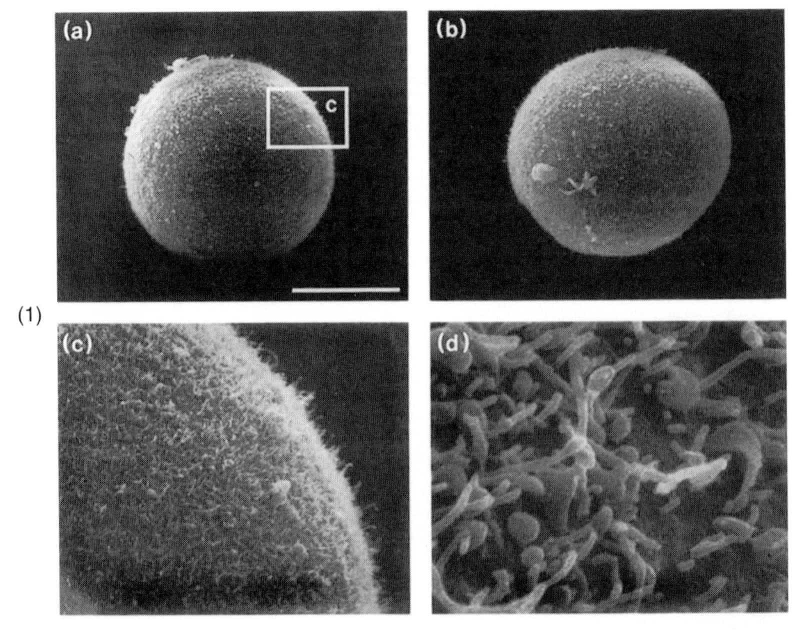

Figure 7.19 The plasma membrane of the human oocyte is organized into microvilli that become shorter and less dense at the zygote stage (1), the four-cell stage (2) and the eight-cell stage (3) (from Dale et al. 1995).

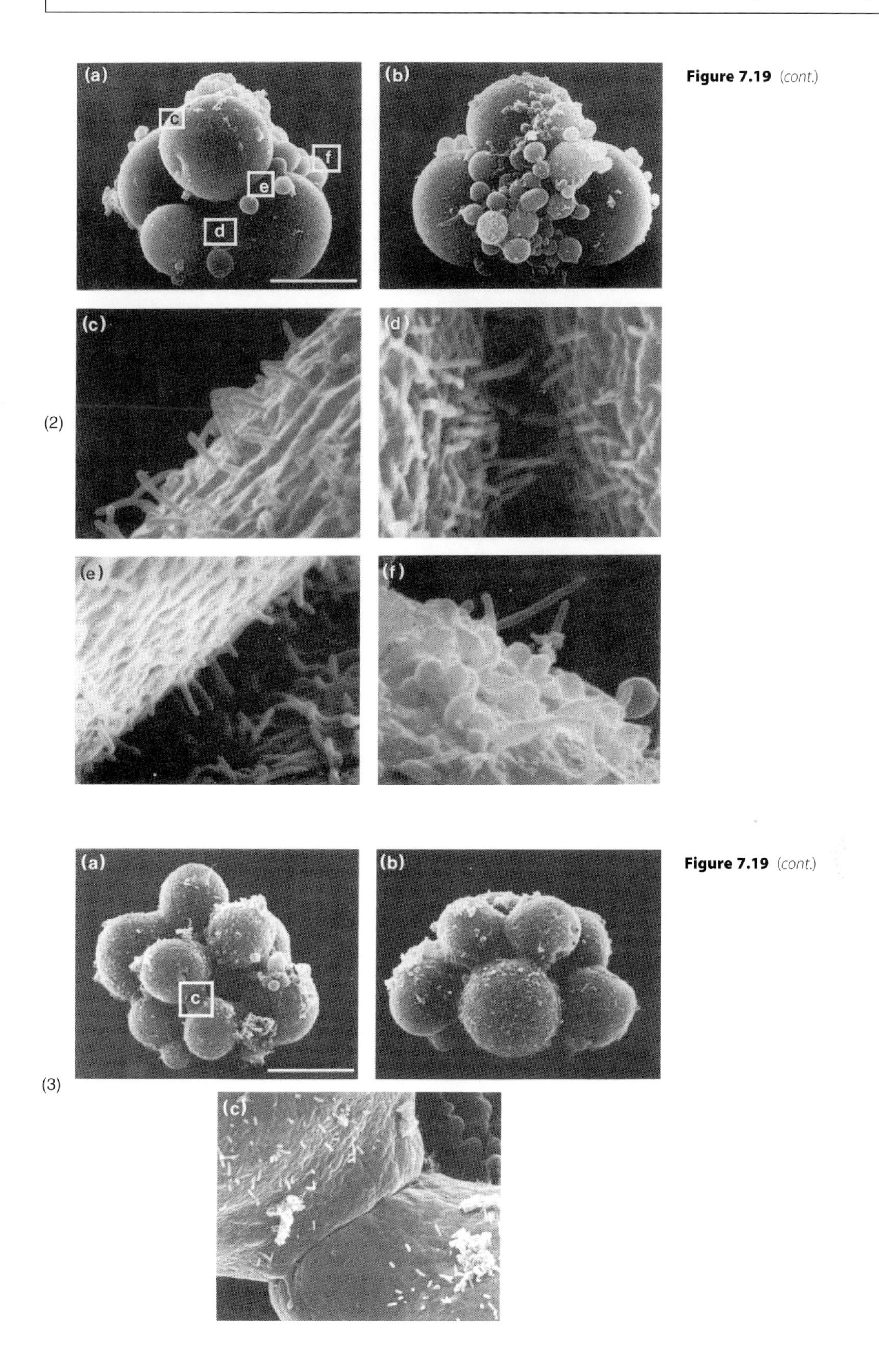

Figure 7.19 *(cont.)*

Figure 7.19 *(cont.)*

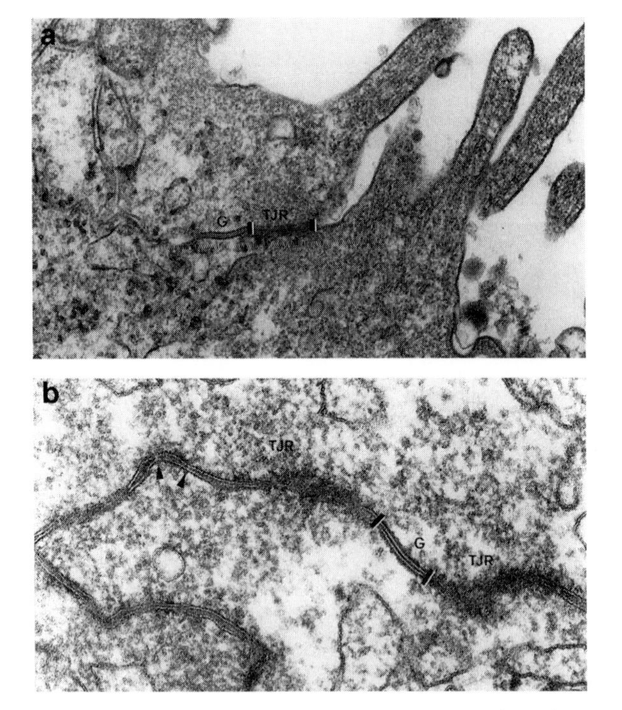

Figure 7.20 In the human embryo, tight junctions are formed at the six- to ten-cell stage, forming an impermeable seal between blastomeres that is essential for the establishment of the fluid filled blastocyst cavity (from Gualtieri et al. 1992).

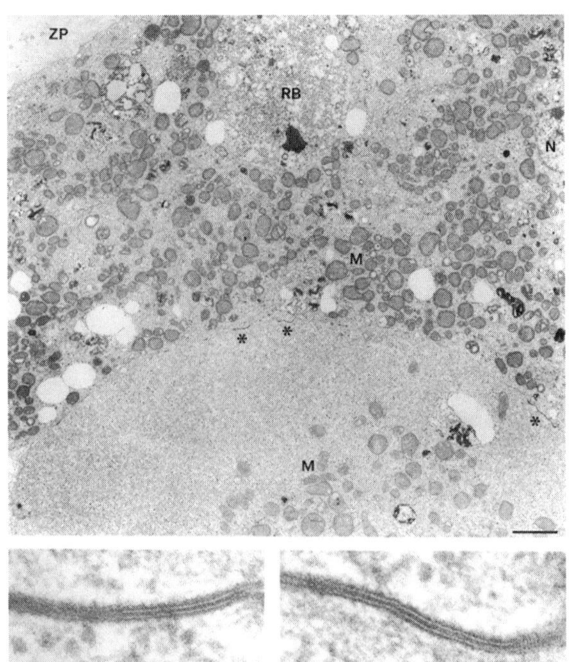

Figure 7.21 Gap junctions in the human embryo are seen in the early blastocyst and allow the transfer of ions and small molecules, such as second messengers between cells. ZP – Zona Pellucida. The gap junctions located by the asterisks are shown at higher magnification in the insets (from Dale et al. 1991).

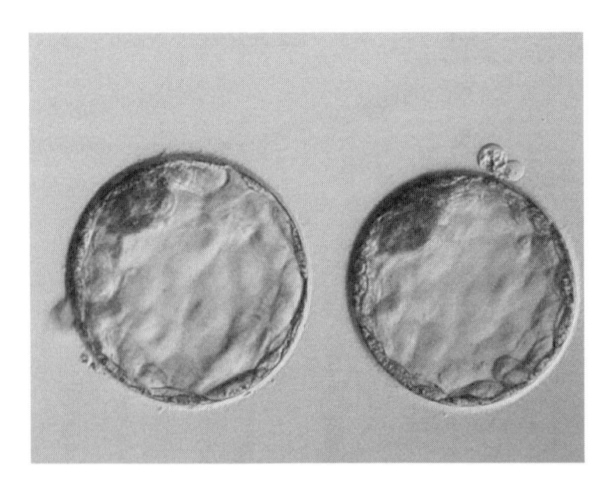

Figure 7.22 Two human blastocysts produced in an in vitro laboratory. Today two healthy young twins (courtesy of Centro Fecondazione Assistita, Naples). (A black-and-white version of this figure will appear in some formats. For the colour version, please refer to the plate section.)

blastocoel. The movement of other ions such as chloride and bicarbonate also contributes to blastocoel formation. Pumping is only possible because the trophectodermal cells become polarized to form an epithelial layer and the tight junctions form a continuous seal between trophectoderm cells and prevent the leakage of small ions out of the blastocoel. The mystical human blastocyst is shown in Figure 7.22.

Chapter 8

Basic Cell Biology

Ion Regulation in Cells

Ion Channels

The difference in ionic constitution of the cytoplasm and intracellular organelles with the extracellular medium is maintained by the hydrophobic lipid bilayer and transmembrane proteins – ion channels and transporters. All cell types, including gametes, survive in this manner, in fact one of the first manifestations of cell death is loss of this ionic homeostasis. It has been estimated that 15–30 per cent of all membrane proteins are involved in transport. Transporters actively transport specific molecules across membranes and may be coupled to an energy source, while channels form a narrow hydrophilic pore, allowing passive movement of small inorganic ions. By generating ionic concentration differences across the lipid bilayer, cell membranes can store potential energy in the form of electrochemical gradients; however, the cell is electrically neutral, e.g. it must contain equal quantities of positive and negative charges. Up to 100×10^6 ions can pass through one open channel each second, 10^5 times greater than the fastest rate of transport mediated by transporters.

Ion transport through channels is passive, i.e. not linked to energy sources, and is often specific for a type of ion. Although channels are essentially hydrophilic pores they are not continuously open, but gated, i.e. they open and close briefly in response to a change in voltage, mechanical stress or the binding of a ligand. Protein phosphorylation and dephosphorylation also regulates the activity of many ion channels. When there is no net flow of ions across the plasma membrane, the resulting transmembrane voltage is called the resting membrane potential, which, using the Nernst equation, may be calculated if the ratio of internal and external ion concentrations are known. The resting potential in most cells depends on the gradient of K^+ across the membrane together with the characteristics of K^+ ion channels. Since there is little Na^+ inside the cell, this has to be balanced by an increase in cations, mainly K^+, which is actively pumped into the cell by the Na^+/K^+ pump and can also move freely in or out through the K^+ leak channels in the plasma membrane. Very few ions adjacent to the plasma membrane (<1nm) actually contribute to the resting potential, and a small flow of ions carries sufficient charge to cause a large change in the membrane potential. Generally, the more permeable the membrane is to a specific ion, the closer the potential will be to the equilibrium potential for that ion. Since 1 microcoulomb of charge (6×10^{12} monovalent ions) per square centimetre of membrane, transferred from one side of the membrane to the other, changes the membrane potential by roughly 1 V, then in a spherical cell of 10 microns diameter about 1/100,000 of the total number of K^+ ions in the cytosol have to flow out to alter the membrane potential by 100 mV.

The plasma membrane of many cells also contain voltage-gated cation channels, which are responsible for depolarizing the plasma membrane, i.e. to a less negative value inside. Voltage-gated Na^+ channels, allow a small amount of Na^+ to enter the cell down its electrochemical gradient and this then depolarizes the membrane further, thereby opening more Na^+ channels, and so on in an auto amplification mode. In an action potential, this may shift the membrane from −70 mV to +50 mV in a fraction of a second. The Na^+ channels then inactivate, and voltage-gated K^+ channels open to return the plasma membrane potential back to its original resting potential. The efflux of K^+ is much more powerful that the influx of Na^+ and quickly drives the membrane back towards the K^+ equilibrium potential of −70 mV. Voltage-gated sodium channels are primarily responsible for action potential propagation in neurons. The channel family consists of nine members. These channels have two subunits, α and β and several membrane-spanning regions. The primary effect of voltage-gated sodium channels in the plasma membrane is a depolarization of the cell through the influx of sodium ions. In contrast to that of sodium channels, the primary effect of the opening of voltage-gated potassium channels is a hyperpolarization of the plasma membrane. This occurs because potassium ions leave the cell cytoplasm, causing a net increase in negative charge in the cell cytoplasm. The role of these channels is to regulate the depolarization caused by voltage-gated sodium channels, causing the cell to repolarize after an action potential. Potassium channels have a tetrameric structure consisting of four subunits. Voltage-gated calcium channels allow the entry of calcium ions into the cell after depolarization. The channel is a complex structure consisting of α_1, $\alpha_2\delta$, β_{1-4} and γ subunits. Voltage-gated calcium channels are commonly involved in muscle contraction, gene expression and neurotransmitter release. There are five common types of voltage-gated calcium channels; L-type, N-type, P/Q type, R-type and T-type. Voltage-gated chloride channels also exist, and play a role in resetting the action potential caused by the opening of other voltage-gated channels.

Cell membranes contain thousands of ion channels, and recording with an intracellular microelectrode allows a qualitative measurement of the membrane potential. To quantify and measure the actual currents underlying these voltage changes, it is necessary to voltage clamp the membrane with a second intracellular microelectrode. In 1976, Neher and Sakman refined a much more superior technique for voltage clamping with a single electrode called patch clamping. This new technique, using a fire-polished micropipette of about 1 micron in diameter, demonstrated current flow through a single channel and took electrophysiology to the molecular level. Breaking the G–ohm seal by suction gives access to the cell interior and enables the researcher to whole cell voltage clamp the cell with a single micropipette. Since that date, many channel types have been classified in a variety of cells. Patch-clamp recording showed that individual voltage-gated Na^+ channels open in an all-or-nothing fashion. A channel opens and closes at random, but when open, the channel always has the same conductance, allowing 1,000 ions to pass per millisecond. Therefore, the total current across the membrane reflects the total number of channels that are open at any one time.

Voltage-gated Na^+, K^+ and Ca^{2+} channels have positively charged amino acids in one of their transmembrane segments that responds to depolarization by opening the channels. Despite their diversity, all these voltage channels belong to a large super family of related proteins. Another gene family embraces Cl^- channels and a class of ligand gated channels activated by ATP. Ligand-gated ion channels are relatively insensitive to the membrane potential and therefore cannot by themselves produce a self-amplifying depolarization. The best example, of a ligand-gated ion channel is the acetylcholine receptor, which was the first channel to be sequenced. The acetylcholine receptor of skeletal muscle is composed of five transmembrane polypeptides encoded by four separate genes, and is non-specific for ion selectivity. Na^+, K^+ and Ca^{2+} may pass through the acetylcholine-gated channel.

Transporters or Pumps

Transporters are long polypeptide chains that cross the lipid bilayer several times and transfer bound solutes across the membrane either passively or actively. Transporters are often called pumps since they are able to 'pump' certain solutes across the membrane against their electrochemical gradients. This active transport is tightly coupled to a source of metabolic energy, such as ATP hydrolysis or an ion gradient. The solute-binding sites are alternately exposed on one side of the membrane and then on the other. Transporters may be categorized into uniporters that move the solute from one side of the membrane to the other, symporters that simultaneously transport two solutes in the same direction, or antiporters that transfer two solutes but in opposite directions.

Antiporters, pH and Ca^{2+} Regulation

Marine and mammalian cells need to maintain their pH around a value of 7.2 for the correct functioning of enzymes. To do this, they employ one or more Na^+-driven antiporters in their plasma membranes, which use the energy stored in the Na^+ gradient to pump out excess H^+. One example is the Na^+/H^+ exchanger, which couples an influx of Na^+ to an efflux of H^+, while another is the Na^+ driven $Cl^-/HCO3^-$ exchanger that couples an influx of Na^+ and $HCO3^-$ to an efflux of Cl^- and H^+. Free cytosolic calcium needs to be maintained at very low levels in all cells. Ca^{2+} transporters that actively pump this ion out of the cell are the Ca^{2+} ATPase and the Na^+/Ca^{2+} exchanger.

The Na^+/K^+ Pump

K^+ is typically 20 times higher inside cells than outside, whereas the reverse is true for Na^+ (Figure 8.1). The

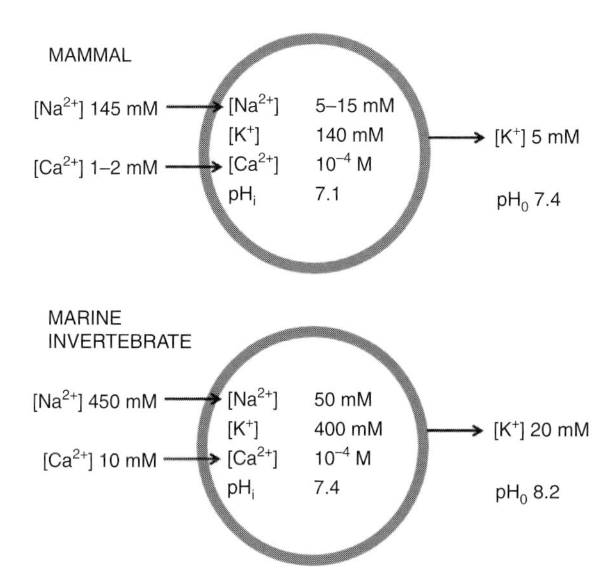

Figure 8.1 The relative concentrations of the major ions inside and outside of cells.

Na^+/K^+ pump maintains these concentration differences, by actively pumping Na^+ out of the cell against its steep electrochemical gradient and pumping K^+ inside. This pump is vital for survival, and it has been estimated that 30 per cent of cellular energy is devoted to its activity. Since the pump is an enzyme, it can work in reverse to produce ATP. The electrochemical gradients for Na^+ and K^+ and the relative concentrations of ATP, ADP and phosphate determine whether ATP is synthesized or Na^+ is pumped out of the cell. This pump is also involved in regulating osmolarity.

Calcium Regulation

Calcium signals may have developed as a side product from the necessity to buffer this ion to very low levels in the cell cytoplasm. In fact, most extracellular fluids contain large quantities of calcium (millimolar range), whereas the cytoplasm of most cells contains minute quantities (nanomolar range). Therefore, calcium ions tend to enter cells by diffusion. Because raised levels of calcium in the cell cytoplasm for extended periods of time are toxic, cells compensate for this net movement of calcium ions both by rendering the plasma membrane of cells highly impermeable to calcium ions, and by elimination of excess calcium from the cytoplasm. So how is calcium buffered in the oocyte cytoplasm? Early models on calcium buffering suggested that the intracellular calcium store was refilled directly from the external milleu through a 'capacitative' mechanism without passing through the cytoplasm. These data were based on the fact that no or little calcium increases were measured with fluorescence dyes when calcium-depleted cells in calcium-free external media were replaced in calcium-containing medium. However, these data were based on the measurement of calcium in the oocyte cytoplasm with fluorescent dyes, which were later found to lose efficiency through compartmentalization within the oocyte cytoplasm. If we add the hypothesis that cytoplasmic calcium pumps are highly efficient at removing calcium from the cytoplasm, we can hypothesize that calcium fluxes into the cytoplasm and is quickly pumped into intracellular stores. Calcium is in fact eliminated from the cytoplasm both through a 110 kDa Ca^{2+}-transport ATPase on the endoplasmic reticulum (ER) membrane, a Na^+/Ca^{2+} exchanger which pumps calcium ions out of the cell and by sequestering into the mitochondrial matrix. In the ER, calcium binds to a variety of proteins, including calsequestrins and calreticulin, and is stored. These proteins, as elucidated from muscle, have a large capacity for calcium binding (about 40–50 moles Ca^{2+}/mole protein), which permit the storage of large amounts of calcium and the rapid release of these ions when necessary. Therefore, two calcium gradients are present within cells: the gradient from the external environment to the cytoplasm, and the gradient between the internal store (ER) and the cytoplasm. The calcium gradients that exist within cells therefore can be utilized as intracellular signals (Figure 8.2).

In order to use calcium gradients within cells as messengers, cells must possess two further mechanisms: first, a mechanism enabling the release of a short burst of calcium into the cytoplasm in response to other signals, and second, a mechanism to 'read' these signals and translate them into cellular signals. Cells achieve these necessities through a combination of receptor-operated calcium channels on the calcium stores and proteins that respond to calcium signals causing a cascade of phosphorylation/dephosphorylation reactions that translate into specific activities.

In order to harness the signalling capacity of calcium gradients within cells, mechanisms of controlling the flux of calcium in response to a specific signal must be present. This is achieved in the form of calcium channels. A variety of calcium channels have been discovered in vertebrate cells, and homologues of these have been shown to exist in all eukaryotic cells

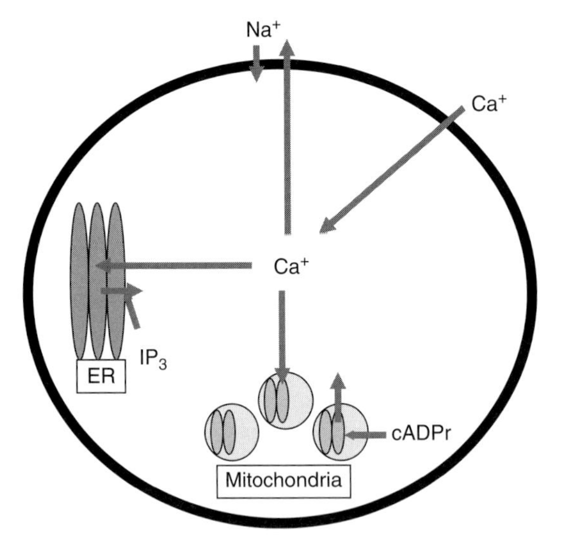

Figure 8.2 Pathways of calcium in eukaryotic cells. Calcium enters the cytoplasm from the extracellular environment down its concentration gradient and is either again expelled into the extracellular environment through a Na^{+}/Ca^{2+} exchanger or sequestered in the cell within the mitochondria or endoplasmic reticulum. These intracellular calcium stores provide sources of calcium for intracellular calcium release. Two pathways are shown: IP_3-sensitive calcium release from the endoplasmic reticulum and CICR from the mitochondria.

tested. The types of calcium channels mainly fall into four distinct groups. Voltage-gated calcium channels were discovered in cardiac and neuronal cells and major groups in this category are L- and T-type voltage-dependent calcium channels. Voltage-dependent calcium channels are found in the oocyte plasma membrane. Receptor-operated calcium channels include the NMDA receptor found in neuronal tissue and the ATP receptor found in smooth muscle. Second messenger-operated calcium channels include the inositol trisphosphate (IP_3) and calcium-induced calcium release (CICR)-activated group of channels. Lastly, calcium channels have been found that are sensitive to physical forces and stretching, these may regulate cell size and response to injury. The IP_3 and CICR receptors are of major interest for fertilization and early development.

The Ryanodine-sensitive calcium release channel is a channel discovered in skeletal muscle due to the sensitivity of this receptor to the plant alkaloid Ryanodine. The receptor has been purified in several laboratories. The receptor monomer is a 450 kDa protein and the receptor as such consists of a tetrameric unit with a Ryanodine-binding site, a Ca^{2+}-release channel and a membrane-spanning domain. The activity of the channel is enhanced by caffeine, adenine nucleotides and calcium itself; whereas ruthenium red or procaine act as inhibitors. The Ryanodine-sensitive calcium channel is thought to be responsible for inositol trisphosphate (IP_3)-insensitive CICR in many systems, suggesting a further sensitivity to calcium itself.

The IP_3-sensitive calcium release channel is a channel found in non-muscle cells and was discovered initially as a channel gated by hormone-ligand interactions on the cell surface. The channel was purified initially from cerebellar Purkinje cells, which are highly enriched in ligand-binding sites. The IP_3 receptor was found to be a tetramer of 260 kDa subunits and such receptors are found in neurons as well as in oocytes of many species. The IP_3 receptor releases calcium in response to IP_3 and related molecules. Heparin is a known inhibitor of IP_3-induced calcium release. Interestingly, the IP_3 receptor – like the Ryanodine channel – shows a bell-type sensitivity to calcium, i.e. the channel is first sensitized, and then desensitized in the presence of increasing amounts of Ca^{2+}. These data suggest that a small amount of calcium release has a positive feedback effect on further calcium release, which eventually stops both through store emptying and channel desensitization. This property of calcium channels helps explain the phenomenon of calcium spiking observed in many cell types.

Basically speaking, oocytes, as other cells, have three calcium release mechanisms. Firstly, the influx of calcium from the external milieu can be regulated through the voltage-dependent calcium channels in the plasma membrane. Second, the production of IP_3 within the cell binds to a receptor-operated calcium channel on the ER, causing the release of calcium from internal stores. Third, CICR describes the mechanism in which calcium itself causes a further release of calcium either by sensitizing the IP_3 receptor to IICR or through the action of a third channel termed the ryanodine receptor.

Inositol 1,4,5-trisphosphate (IP_3) was found to be a potent non-muscle calcium-releasing agent through the work of Berridge and collaborators in the 1980s. The initial importance of IP_3 was that this molecule was the link between hormone-receptor interaction and the cell response. IP_3-induced calcium release was subsequently found to be a mechanism for calcium release in many species in response to both external and internal stimuli. IP_3 is produced through hydrolysis of polyphosphoinositides on the plasma membrane by phospholipase C (PLC). The activation of

PLC appears to be achieved through a GTP-dependent (G-protein) mechanism.

Calcium-induced calcium release was initially discovered in muscle cells through the use of Ryanodine. Here, this product causes a massive muscle contraction by binding to the 'Ryanodine receptor' and causing the release of calcium. Interestingly, the natural ligand for the Ryanodine receptor, cADPr, in non-muscle cells was only discovered in 1989, despite the great interest in these receptors. cADPr is a metabolite of NAD^+ and has now been shown to be an active calcium-releasing metabolite in many species. However, other calcium-releasing metabolites also exist (such as $NAADP^+$, derived from $NADP^+$), and have been shown in turn to possess calcium-releasing properties, suggesting that metabolites of nicotinamide form a family of calcium-releasing second messengers. cADPr is produced through the activity of adenosine diphosphate-ribosyl cyclases which are in turn regulated by levels of cyclic guanosine monophosphate (cGMP) and may be produced after hormonal stimulation in some cell types such as pancreatic β-cells. The potent calcium-releasing activity of cADPr was discovered initially in sea urchin oocytes.

Calcium transients – or spikes – in the cell cytoplasm – were first measured in the 1970s with the use of calcium-sensitive aequorin proteins – that released light in the presence of calcium, and calcium-sensitive fluorescence dyes such as fura-2 and fluo 3. These proteins, although excellent in the demonstration of transient increases in intracellular calcium – gave little information on the properties of these spikes. It was not until the introduction of two-dimensional cell imaging such as photo-imaging detectors and confocal microscopy that the properties of calcium peaks observed in many cell types were observed. Calcium spikes are now known to either remain localized to distinct regions of the cell cytoplasm or cross the cytoplasm in the form of a wave. Waves of calcium release are common to many cell types and especially common in oocytes of many species at fertilization. Interestingly, in some cells calcium increases simultaneously throughout the whole cytoplasm. The mechanisms for the formation of calcium waves or the simultaneous increase in calcium appear distinct for each species. However, one common feature underlies the calcium transients – they are produced by either IP_3-induced mechanisms or CICR-induced mechanisms – or both.

Calcium spikes regulate cell activities through the action of two major proteins: calmodulin and calcium/calmodulin-dependent protein kinase II (CaMKII). Calmodulin is a 16 kDa ubiquitous calcium binding protein that mediates a host of cell processes in many cell types in response to a calcium signal. Calmodulin is in fact the primary messenger of calcium signals since most cellular proteins are unable to bind to calcium itself. Calmodulin undergoes a conformational change upon binding calcium and possesses four sites for calcium binding. The occupation of all sites for Ca^{2+} is not necessary for all the functions of calmodulin, suggesting that different levels of cytoplasmic calcium can activate diverse processes. The relevance of calmodulin during the cell cycle is inferred from its' localization: calmodulin is localized throughout the cytoplasm during interphase but migrates to the mitotic apparatus during M-phase.

The Calcium/calmodulin-dependent protein kinases are a series of serine/threonine protein kinases activated by calmodulin. The kinases are oligomeric proteins with diverse subunits of approximately 50–60 kDa forming a complex of between 300–600 kDa depending on cell type. The kinase is characterized by a 'memory' effect, i.e. the activation of the kinase supercedes the presence of Ca^{2+}/calmodulin. Two types of CaM kinase exist: specialized CaM kinases such as Myosin light chain kinase (MLCK) involved in muscle contraction, and multifunctional CaM kinases such as CaM kinase II. CaM kinase II is involved in a host of cellular processes and is relatively non-specific for substrates, leading to the postulation of how Cam kinase II can organize specific cell processes.

How then can a cell trigger specific activities through a calcium spike? Probably, the calcium spike at fertilization is designed to be just as shown – a large, non-specific signal that causes several major effects including the degradation of cell cycle blocks, release of cortical granules, upregulation of metabolism, de-condensation of the sperm head and the activation of the programme of development. However, calcium spikes are also often triggers of specific mechanisms. Cells appear often to achieve these specific effects through localization of calcium signals. Localized calcium spikes appear in many classes of cells such as neurons, pancreatic cells and embryos: a good example of localized calcium waves are observed in sea urchin oocytes after activation.

The Cell Cycle

During mitosis, a regular growth phase (interphase, G_1) in which the cell volume increases is followed by a phase of replication in which an exact copy of the entire genome of a cell is made (S-phase). This phase is then followed after a second growth phase (G_2) by the separation of these copies and division of the cell into two daughter cells, which contain two sets of chromosomes as in the parent cell (M-phase). M-phase consists of six stages. During prophase, chromosomes condense within the nuclear envelope. Prometaphase describes the stage of nuclear envelope breakdown and development of the mitotic spindle. The completion of prometaphase is termed metaphase (where chromosomes are correctly aligned along the mitotic spindle). Anaphase follows metaphase, and during this stage sister chromatids separate and move to opposite poles of the mitotic apparatus. When anaphase is complete, the mitotic apparatus starts to disintegrate and the cell initiates cleavage (telophase). The end of telophase is termed cytokinesis, and consists in the formation of two separate daughter cells in interphase with a reformed nuclear envelope. The first growth phase is termed G_1 (between M-phase and S-phase), and a shorter second phase termed G_2 occurs between S-phase and M-phase. In somatic cells, a state of quiescence or cell cycle block in response to a specific physiological state of the cell is described as the G_0 phase of the cell cycle. However, G_0 differs from meiotic blocks in terms of cell cycle regulation and the activity of the key kinases that maintain the arrest. Meiosis is also slightly different to mitosis in this respect in that a single round of genome duplication (S-phase) is followed by two rounds of division (M-phase) without chromosome replication. In this way, four daughter cells are formed, each with half the chromosome number (haploid). The daughter cells are termed 'gametes' and are highly specialized cells in which the unique purpose is to unite with another gamete to form a new, distinct diploid cell (Figure 8.3).

Checkpoint Controls

Each cell cycle is controlled at critical points in development. These points are termed 'checkpoints'. During mitosis, checkpoints occur at the initiation of DNA replication (to prevent this process occurring before the cell volume has reached a certain size, i.e. before the cell has assimilated enough nutrients to complete the cell cycle), at the beginning of cleavage (to prevent this process occurring before DNA replication is complete or in the presence of damaged DNA) and prior to cell cleavage (to prevent cytokinesis occurring before paired chromosomes are correctly aligned on the mitotic spindle). Although the DNA replication associated checkpoints appear molecular in design, interestingly, micromanipulation experiments in spermatocytes show that tension on the spindle generated by attached homologues acts as the mitosis-exit checkpoint. If this tension is eliminated by experimental manipulation, anaphase is prevented. Rigid controls are in place during mitosis in order to ensure that cell division occurs correctly, and when these controls are bypassed, cell division can become uncontrolled such as in the case of tumours.

During meiotic division, cell cycle checkpoints are in place as in mitosis. However, the expression of these is altered with respect to mitosis. During oocyte growth, a checkpoint ensures that the genome is copied completely prior to G_2, but during oocyte maturation this checkpoint is switched off because no DNA synthesis occurs between metaphase 1 and 2. Furthermore, during G_2 of meiosis, a unique process occurs. This process – termed recombination, causes the mixing of genes between chromosomes. Recombination occurs after the pairing of homologous chromosomes during prophase 1 and consists in the exchange of segments of DNA between sister chromosomes. Recombination must be completed before the beginning of cell division so that a correct segregation of homologous chromosomes is obtained. A checkpoint specific to meiotic cells ensures that meiosis 1 does not progress until all DNA strand breaks (essential for the mixing of genes during recombination) are repaired and paired chromosomes are correctly attached to the spindle. This resembles the spindle-assembly checkpoint of mitotic cells.

Although checkpoint controls are rigidly adhered to during mitosis, evidence suggests that during meiosis the controls are weaker. The reasons for the loosening of checkpoint controls during meiosis are unclear, but may reside in the necessity to enable embryo development to occur rapidly in order to permit embryo implantation before embryo resources are depleted. In fact, the oocyte is the unique store of nutrients in pre-implantation embryos and embryos depend almost entirely on this store until implantation (or gastrulation in lower species). It could be assumed that the relaxing of checkpoint controls was evolutionarily necessary to enable rapid embryo development, the wastage of a large quantity of

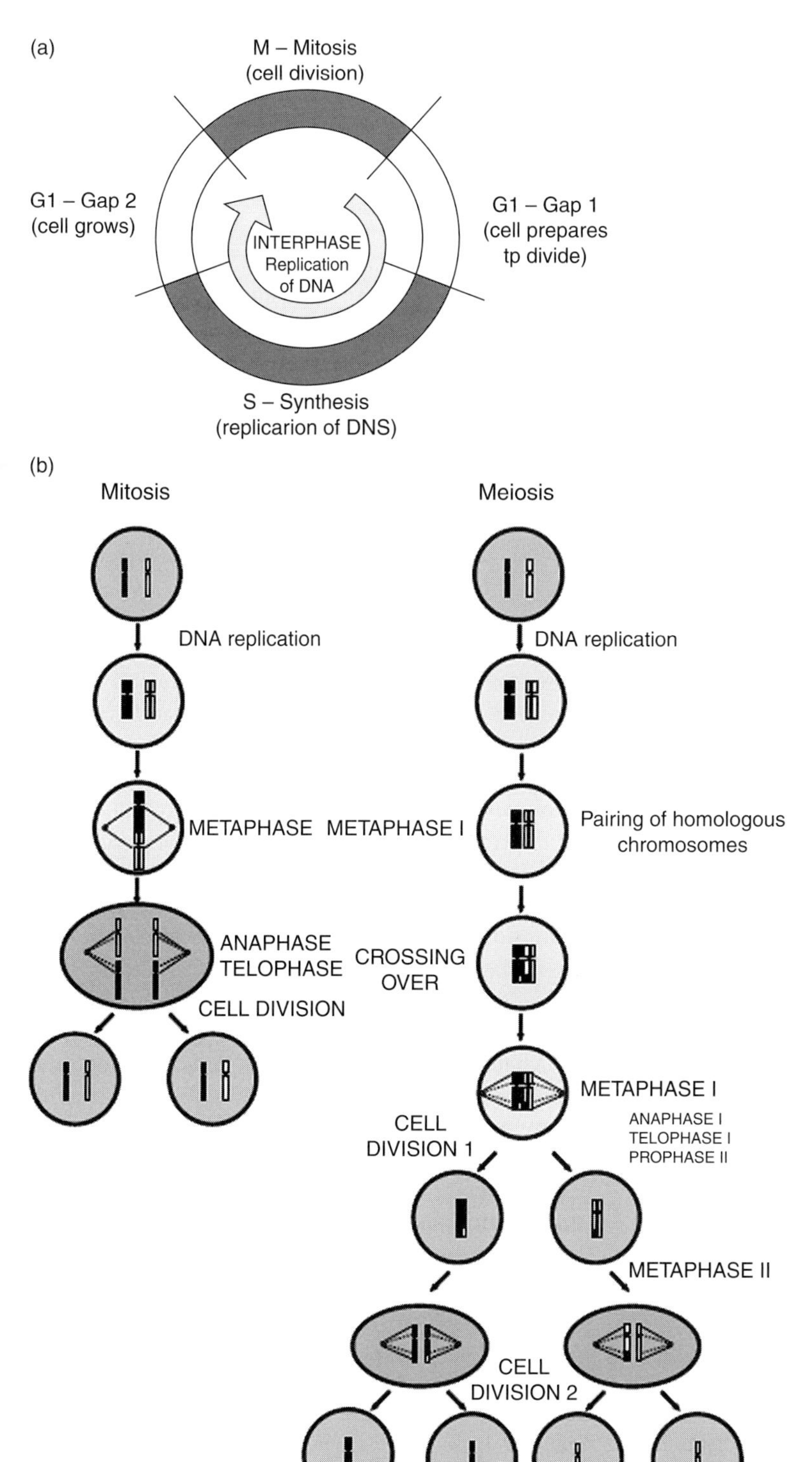

Figure 8.3 **(a)** A scheme of the cell cycle and **(b)** mitosis vs. meiosis.

embryos through mistakes in meiosis and mitosis being the side product of this process.

Meiosis and Mitosis: Intracellular Mechanisms

Despite the differences between mitosis and meiosis, the underlying mechanisms are the same for all species. In the early 1970s, oocyte cytoplasmic transfer experiments led Masui and colleagues to discover that maturing amphibian oocytes produce a factor that causes them to resume meiosis. This factor was termed maturation promoting factor, or MPF. MPF is now used to represent M-phase promoting factor, reflecting its involvement in both meiosis and mitosis. The molecular mechanisms for the activity of this factor were determined from experiments both in frog oocytes and in yeast molecular biology. The major protein involved in MPF is termed $p34^{cdc2}$. $p34^{cdc2}$ is a serine/threonine kinase that phosphorylates a host of cellular proteins, bringing about the condensation of chromosomes, the dissolution of the nuclear envelope and the formation of the mitotic apparatus, hence the transition from G_2 to M-phase. The destruction of MPF activity is now known to underlay the mechanism for the completion of mitosis (i.e. the M-phase to G_1-phase transition).

Regulation of MPF

The regulation of MPF is tightly linked to cell cycle progression and checkpoint controls. The mechanisms of regulation are now fairly well understood and delineated, through work mainly on yeast and frog oocytes. MPF exists as a complex with a regulatory subunit family of proteins termed 'cyclins'. The cyclins were discovered as proteins whose quantity increased during the G_1 and S-phase of the cell cycle, and was rapidly destroyed during the M to G_1 transition. The presence of the cyclins, particularly cyclin B, was necessary for high MPF activity, and the destruction of these proteins correlated with the destruction of MPF activity. The cyclins have no activity per se and are therefore regarded as the regulatory subunit of MPF. The major cyclin is termed cyclin B_2. Cyclin protein levels build up during G_1 and S-phase, and are rapidly destroyed during M-phase – hence the term 'cyclins'. If the simple presence of the cyclins was sufficient for entry into M-phase, there would be little control over the cell cycle as cells could enter into M-phase as soon as sufficient cyclin protein was present, which is in fact well before entry into M-phase. In fact, cyclins are necessary for entry into M-phase, but not sufficient. Apart from the necessity of the cyclins, MPF is regulated in itself by a cascade of phosphorylation/dephosphorylation reactions. Fission yeast $p34^{cdc2}$ is phosphorylated on two sites prior to M-phase: Tyr 15 and Thr 167, although Tyr 15 is the amino acid residue vital for M-phase progression. Phosphorylation of Tyr 15 appears to be maintained by the activity of the weel protein kinase (yeast) and its homologues. Tyr 15 is abruptly dephosphorylated prior to the G_2/M-phase transition. In fact, Tyr dephosphorylation appears to be required for activation of the MPF complex. A protein phosphatase, $p80^{cdc25}$ (yeast) and its homologues appears to be required for this process. The production of $p80^{cdc25}$ appears to be inhibited by the presence of unreplicated DNA, providing a mechanism for the prevention of M-phase entry prior to the completion of S-phase. The mechanism suggests that at the M-phase transition, the overproduction of $p80^{cdc25}$ causes net dephosphorylation of Tyr 15- driving MPF towards activation, although this model appears still to lack precision and other mechanisms have been proposed. One such mechanism suggests that intracellular signals such as calcium could provide the ultimate trigger for M-phase entry. In fact, calcium signals are universally observed at fertilization (in which the cell cycle checkpoint is bypassed by fertilization) and have been observed during mitotic cell cycle transitions in many species. Calcium signals activate calmodulin, and the regulation of cell cycle proteins by calmodulin has been suggested as the ultimate trigger for passage through the cell cycle transitions.

After the activation of MPF and the entry into M-phase, MPF activity is abruptly destroyed. This process permits the cell cycle to progress through anaphase and cell division to occur. This process appears to depend on the correct alignment of chromosomes on the mitotic apparatus. As stated above, micromanipulation experiments in spermatocytes show that tension on the spindle generated by attached homologues acts as the mitosis-exit checkpoint. If this tension is eliminated by experimental manipulation, anaphase is prevented. MPF activity is destroyed by the protease-dependent destruction of the cyclin subunit of MPF. The mechanism appears to involve ubiquitination of cyclins followed by degeneration of these by the 26S proteasome. Selection of substrates for degradation is performed by an anaphase-promoting complex/cyclosome (APC/C), which has ubiquitin ligase activity. APC/C activity is inhibited by the proteins Mad2, which prevent anaphase in the absence of attached chromosomes to spindles. As for the mechanism controlling M-phase entry, the precise trigger of anaphase entry is not yet clear, and cell signals such as calcium have also been proposed. Again, evidence derives from the fertilization of oocytes arrested at anaphase, as well as the observation of calcium signals during anaphase progression in many species.

Regulation of Meiosis: CSF

As stated previously, meiosis is a permutation of the mitotic cell cycle in that, although the cell undergoes the normal G_2 to M-phase transition, a second M-phase occurs without any intervening DNA replication (S-phase), resulting in four daughter cells with half the chromosome constitution of the parents.

Apart from the inhibition of S-phase during meiotic divisions, meiosis is distinct from mitosis in that oocytes are characterized by a further block during meiosis to await fertilization. The mechanisms of the second meiotic block are not common to all animals, for example the block in the sea urchin oocyte is due to the acidification of the cytoplasm, and fertilization reverses this process. Many animals though are characterized by the presence of a second protein complex which actively arrests meiosis during M-phase by stabilizing the active MPF complex, causing the failure of the oocyte to separate chromosomes and proceed to telophase. This second protein complex, termed cytostatic factor (CSF), was discovered through experiments on frog oocytes by Masui and colleagues and is, like MPF, a serine/threonine kinase. Although the biological activity of CSF has been largely characterized, its biochemical components and mechanism of action remain elusive. It is generally accepted that the main components of CSF include the product of the proto-oncogene *c-mos*, mitogen associated protein kinase (MAPK), and possibly cdk2 kinase. The *c-mos* proto-oncogene product, Mos protein, expressed in female germ cells, has a specific role in the meiotic cell cycle. Mos is a serine/threonine kinase, and in oocytes of several species, its activity increases at the time of GVBD, and remains high until M-II. Inhibition of Mos expression prevents GVBD in frog oocytes. In oocytes that block during meiosis 2, Mos appears to control this block by maintaining MPF at a high level of activity through the prevention of cyclin degradation, probably through the phosphorylation and subsequent activation of MAP kinase. However, although fertilization in species arrested at metaphase 2 induces the inactivation of MPF and CSF, CSF activity declines more slowly than MPF, suggesting independent mechanisms for the exit from metaphase 2. Interestingly, fertilization in oocytes that arrest at other points in meiosis causes the activation of CSF. In clams (fertilization at the GV stage) CSF is inactive prior to fertilization, increases during meiotic progression and declines during metaphase 2. In ascidians (fertilization during metaphase 1) CSF activity is present in unfertilized oocytes, but rapidly increases only to decline again during metaphase 2. Therefore, CSF activity appears to be required during meiosis, and peaks during metaphase 2 even in oocytes in which no arrest occurs at this stage. CSF activity may therefore be the factor that inhibits S-phase during meiotic progression, determining the reduction division characteristic of meiosis.

Mos is a potent activator of the MAP kinase pathway, and MAPK may phosphorylate some of the substrates modified by the active MPF, preventing the oocyte from entering interphase between metaphase 1 and 2. The major substrate of MAPK is now known to be the protein kinase p90Rsk, which causes the inhibition of cyclin degeneration by the APC/C complex. It therefore maintains chromosome condensation and is involved in the reorganization of microtubules that leads to spindle formation at metaphase 1, contributing to the stabilization of the spindle at metaphase 2. It therefore mediates a spindle assembly checkpoint.

Oocytes of all species known are released from cell cycle arrest by an increase in intracellular calcium after fertilization. This calcium increase appears to inactivate Mos through protease activity, leading to the decline in MAP kinase activity. However, the calcium increase also causes the activation of a multifunctional Ca^{2+}/calmodulin-dependent protein kinase, which activates the ubiquitin-dependent cyclin degradation pathway independently of CSF activity. This provides for a mechanism for the rapid release of oocytes from the meiotic block, prior to the Ca^{2+}-dependent inactivation of CSF.

Resumption of Meiosis

During mammalian oogenesis, the oocytes are incapable of activating the molecular machinery for the completion of meiosis in preantral follicles. At the time of antrum formation, the oocyte becomes intrinsically capable of resuming maturation, but maturation arrest is maintained by the integrity of the follicle wall. Minor alterations in the architecture of the follicle wall, the cells surrounding the oocyte or the connections between the cumulus oocyte complex and the mural granulosa trigger meiosis. The pre-ovulatory LH surge removes the inhibitory influence of the follicle wall, and replaces it with a stimulating one that reactivates meiotic progression with a complex cascade of events leading to maturation and ovulation. LH receptors are present only on the somatic cells of the follicle, not on the oolemma – and therefore the somatic cells must send a second message signal to the oocyte; a variety of messages are produced which are transmitted to the oocyte in a specific sequential order. Before the LH surge, the oocytes are 'on standby' in the P-I to M-phase transition, and this appears to be maintained by the diffusion of cAMP from the cumulus to the oocyte. LH alters gap junctions, and the flux of inhibitory signals to the oocyte is interrupted, thus causing a decrease in cAMP levels. It has been hypothesized that meiotic arrest is mediated by a cAMP-dependent protein kinase substrate, which undergoes dephosphorylation when cAMP levels decrease. Although the exact substrates have not been identified, it is generally accepted that progression through the meiotic cell cycle is regulated by a series of protein kinases and phosphatases. It is currently thought that Mos translation is negatively regulated by a protein kinase A (PKA)-mediated action of cAMP and is dependent on an active MPF.

cAMP may serve as the physiological inhibitor involved in the maintenance of meiotic arrest; post-ovulatory mature oocytes contain lower levels of cAMP than follicular immature oocytes. Early fusion studies with *Xenopus* oocytes demonstrated that cAMP inhibits MPF activation. Once MPF has been activated, its ability to induce the transition of nuclei to metaphase is no longer sensitive to cAMP, and more recent studies suggest that cAMP regulates post-translational modification of the kinase protein in meiotic prophase. However, it has also been shown that the maturation-associated decrease in intracellular cAMP is essential, but not sufficient, for re-initiation of meiosis, and an activated cAMP-dependent protein kinase (PKA) mediates the negative action of cAMP on resumption of meiosis. The biological activity of MPF seems to be that of histone Hl phosphorylation; it is postulated that a phosphorylated histone Hl alters nucleosome packing, and may contribute to chromosome condensation. In addition, further experiments with the purified kinase subunit suggest a possible role in metaphase on transcription or translation regulation, microfilament rearrangement, reorganization of the intermediate filament network, and nuclear disassembly.

The resumption of oocyte maturation through the action of LH suggests a molecular link between the two processes. Apart from the decrease in cAMP levels triggered by the action of LH, current evidence suggests other activities of this hormone. One such activity could be in the stimulation of molecules such as meiosis activating sterols (MAS), described first by Byskov and co-authors in the 1990s. Although MAS were sufficient to cause GVBD in frog oocytes denuded of cumulus cells, later evidence suggested that MAS were not necessary for oocyte maturation. Furthermore, the kinetics of meiosis re-initiation after MAS addition were slower than after LH stimulation. Therefore, MAS factors may represent one of a series of activators of meiosis induced by LH.

Metabolism

The production of energy (in the form of ATP) within the cell is termed respiration and occurs through two distinct but interrelated routes: aerobic (mitochondrial) and anaerobic (cytoplasmic) respiration. These two routes share the same initial pathway – termed glycolysis (or 'lysis' of glucose). In this pathway, glucose is converted to pyruvate through a series of reactions, which requires no oxygen and produces a small amount of energy. Consequent to the completion of this stage, pyruvate can enter the mitochondria and be converted to CO_2 + H_2O through the highly efficient energy-producing pathway termed aerobic respiration or oxidative phosphorylation (requiring oxygen but produces ΔG=-686kcal/mol), or be converted within the cytoplasm to lactic acid through anaerobic respiration (not oxygen requiring but produces far less ATP, ΔG=-47kcal/mol; 14.6-fold less than oxidative phosphorylation). One point to remember when visualizing energy production in cells is that the two pathways are not mutually exclusive, i.e. both pathways are always active in cells. Probably, as in muscle cells during exercise, aerobic respiration continues at a rate dependent on the oxygen supply and anaerobic respiration completes the energy requirement when aerobic respiration is insufficient. Whether the source of pyruvate for energy production derives from the breakdown of proteins, lipids or other carbohydrates, the question in mammalian embryo development has always been whether pre-implantation embryos utilize aerobic or anaerobic mechanisms for energy production. In fact, the precise metabolic needs of mammalian embryos during pre-implantation development have been notoriously difficult to define. This is because the mammalian embryo is highly flexible in its culture requirements, and can develop normally in a wide range of culture conditions. The current evidence suggests that the respiratory pathways of mammalian reproductive cells may be slightly different between oogenesis, pre-implantation embryo development and foetal development. During oogenesis, oxidative phorsphorylation appears to predominate, although it must be stated here that the metabolism referred to probably originates in the follicle cells surrounding the oocyte. The stages of oogenesis prior to the LH surge are hard to experimentally manipulate without adequate in vitro culture systems, and in this respect little information on the metabolic requirement of the developing follicle has been collected at that stage. However, ultrasound correlations between the state of perifollicular vascularity of developing follicles, measurements of the dissolved oxygen content of follicular fluid, and indicators of the quality of oocytes retrieved from these follicles suggests that mitochondrial metabolism plays an important role in oogenesis. After the LH surge, the rate of oxygen consumption, a measure of mitochondrial metabolism, increases dramatically in the isolated oocyte. Furthermore, the use of iodoacetate to block anaerobic respiration does not inhibit oocyte maturation, whereas oocyte maturation is blocked when aerobic respiration is suppressed.

The experimental measurement and manipulation of oocytes and embryos is facilitated after oocyte maturation since the oocyte leaves the follicle and can be transferred to in vitro systems with ease. Therefore, data on the metabolic requirements of oocytes and embryos after ovulation is abundant. After fertilization in mammalian embryos, measurements of the relative levels of aerobic and anaerobic respiration suggest that both pathways are active. Mammalian oocytes and pre-implantation embryos prefer pyruvate as an energy source. The preference for pyruvate however does not eliminate neither aerobic nor anaerobic respiration as a possible route to ATP production. Lactic acid is present in fluid sampled from human tubules, and is also produced during mammalian embryo development, suggesting a component of anaerobic respiration in the ATP-generating mechanism. Interestingly, mitochondria within the pre-implantation embryo do not replicate, and are characterized by poorly formed cristae, suggesting that the level of mitochondrial respiration is low during this stage. However, this does not exclude a component of mitochondrial respiration in the energy production capacity of the developing embryo. Physiologically, aerobic respiration may occur at low levels during pre-implantation embryo development because ovulation causes the detachment of the oocyte from the oxygen-enriched follicular environment and the passage of the pre-implantation embryo through the feasibly less oxygen-enriched environment of the fallopian tubes. However, we must remember that aerobic respiration is highly efficient and can occur 14.6-fold less than anaerobic respiration per mole of glucose used, therefore it is feasible that a low level of aerobic respiration will be detected under all circumstances. Evidence from mice suggests that embryos can develop and even implant adequately even in the absence of mitochondrial respiration, suggesting that mitochondrial respiration is not necessary for this phase of embryo development. However, it must be again stated that with adequate energy supply through anaerobic respiration, mitochondrial respiration is obviously not necessary. Aerobic respiration appears to be upregulated at blastocyst development because the mitochondrial cristae become more compact, mitochondrial replication is initiated and the utilization of glucose as a carbohydrate substrate increases at this stage. These data suggest that the intracellular energy supply during both oocyte maturation and embryo implantation is achieved mainly through aerobic respiration, but a large component of anaerobic respiration is present during mammalian pre-implantation development.

Mitochondria are elliptical organelles measuring 0.5–1 μm and are most abundant in metabolically active cells. They have a smooth outer membrane and a highly specialized inner membrane organized into folds or cristae to increase its surface area. The inner membrane contains proteins responsible for the oxidation reactions of the respiratory chain, ATP synthetase and transport proteins for the passage of metabolites into and out of the matrix. The outer membrane is permeable to all molecules up to 10,000 daltons. The matrix contains hundreds of enzymes, including those for the oxidation of pyruvate and fatty acids and for the citric acid cycle (Krebs cycle). Here are also located several copies of the mitochondrial DNA genome, special mitochondrial ribosomes and tRNA. Somatic cells contain 40–1,000 mitochondria, spermatozoa 1–100, while oocytes may contain from 100,000 to 1,000,000. Mitochondria are not static and move around the cytoplasm in association with microtubules. In some cells they appear as long chains along specific microtubules while in spermatozoa they become concentrated in the mid-piece and are fundamental for movement of the flagellum. In human oocytes, the mitochondria cluster from the GV stage onwards polarizing in the mature M11 oocyte providing localized energy sources (Figure 8.4).

In most animals, mitochondrial DNA is maternally transmitted, however there are exceptions. In the mollusc *Mytlilus,* there are two mitochondrial genomes, F found in oocytes and M found in the spermatozoa. The F mitochondria are transmitted by females to both male and female offspring, whereas the M mitochondria are only transmitted to males.

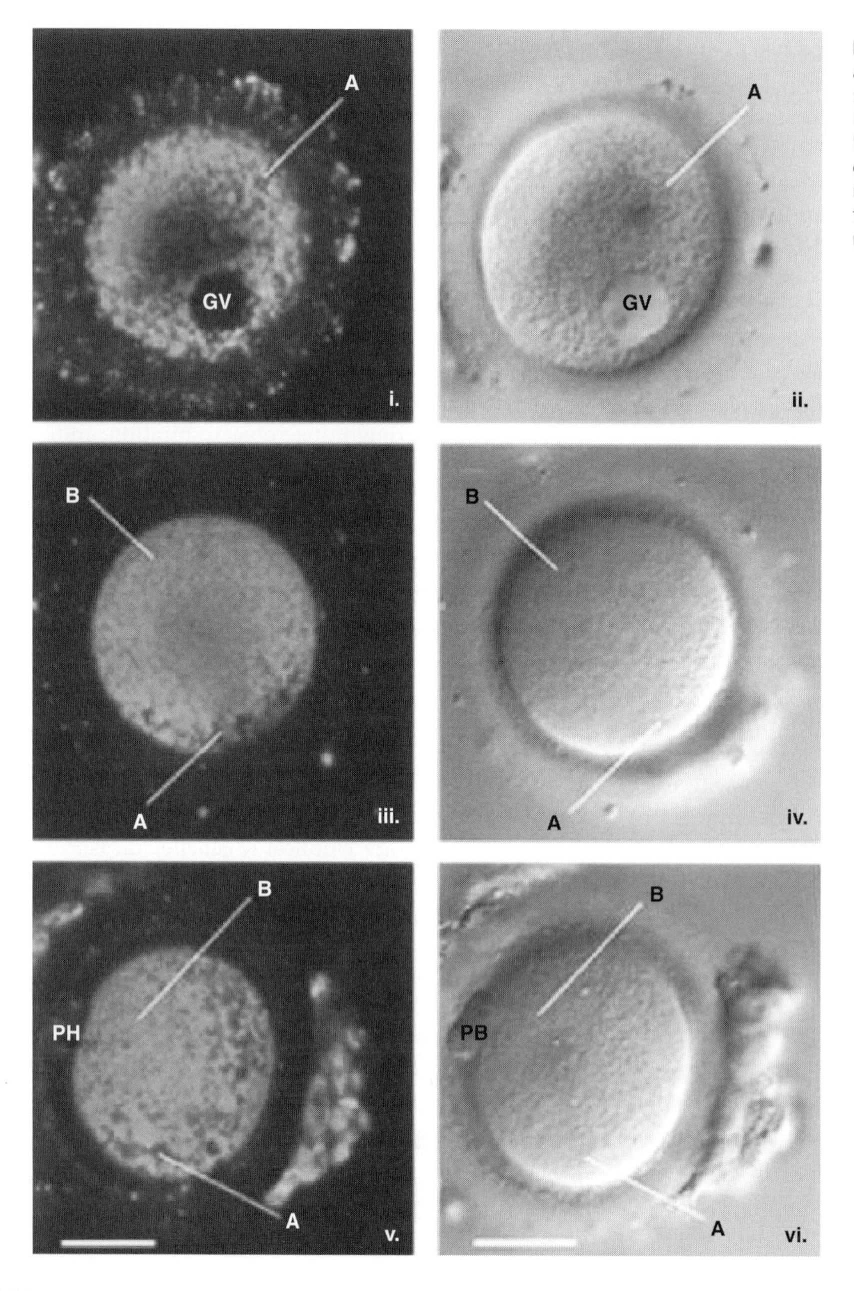

Figure 8.4 Patterns of mitochondrial aggregation in the human oocyte during maturation (from Wilding et al. 2001, Human Reproduction, 16, pp. 909–917). From the top germinal vesicle stage, centre metaphase 1 stage, bottom mature metaphase 2 stage. Frames to the left are in fluorescence showing mitochondrial activity. The bar is 50 µm.

References

Books and Chapters

Adiyodi, K.G., and Adiyodi, R.G. (eds.) (1983) *Reproductive Biology of Invertebrates. Volumes 1-4.* John Wiley, Chichester.

Adiyodi, K.G., and Adiyodi, R.G. (eds.) (1990) *Reproductive Biology of Invertebrates. Volumes 1-4.* John Wiley, Chichester.

Auerbach, L. (1874) *Organologische Studien.* Breslau.

Austin, C. (1965) *Fertilization.* Prentice Hall, Englewood Cliffs.

Austin, C.T., and Short, R.V. (1972) *Reproduction in Mammals.* Cambridge University Press, Cambridge.

Balinsky, B. (1965) *An Introduction to Embryology.* W.B. Saunders, Philadelphia.

Bodmer, C.W. (1968) *Modern Embryology.* Holt, Rinehart and Winston, New York.

Cagnon, C. (ed.) (1999) *The Male Gamete: From Basic Science to Clinical Applications.* Cache River Press, Vienna.

Dale, B. (1983) *Fertilization in Animals.* Edward Arnold, London.

Dale, B. (1996) Fertilization. In: (Greger, R. and Windhorst, U., eds.) *Comprehensive Human Physiology.* Springer Verlag, Berlin, pp. 2265–2276.

De Jonge, C., and Barratt, C. (2006) *The Sperm Cell.* Cambridge University Press, Cambridge.

Edwards, R.G. (1982) *Conception in the Human Female.* Academic Press, New York.

Elder, K., and Dale, B. (2011) *In Vitro Fertilization.* Third edition. Cambridge University Press, Cambridge.

Gordon, N.K., and Gordon, R. (2017) *Embryogenesis Explained.* World Scientific, Singapore.

Hardy, D. (ed.) (2002) *Fertilization.* Academic Press, San Diego.

Huang, Z., and Wells, D. (2011) Molecular aspects of follicular development. In: (Donnez, J. and Kim, S. eds.) *Principles and Practice of Fertility Preservation.* Cambridge University Press, Cambridge, pp. 114–128.

Jamieson, B. (1981) *The Ultrastructure of the Oligochaeta.* Academic Press, London.

Jamieson, B. (1999) Spermatozoal phylogeny of the vertebrata. In: (Claude, C., ed.) *The Male Gamete.* Cache River Press, Vienna, USA, pp. 303–322.

Jamieson, B. (ed.) (2006) *Reproductive Biology and Phylogeny of Annelida.* Science Publishers, Enfield.

Jamieson, B. (ed.) (2007) *Reproductive Biology and Phylogeny of Birds.* Science Publishers, Enfield.

Johnson, M. (2007) *Essential Reproduction.* Sixth edition. Blackwell Scientific Publications, Malden.

Johnson, M., and Everitt, B. (1990) *Essential Reproduction.* Blackwell Scientific Publications, Oxford.

Just, E.E. (1939) *The Biology of the Cell Surface.* Blakiston, Philadelphia.

Lillie, F. (1916) The history of the fertilization problem. *Science* **43**:39–53.

Longo, F. (1987) *Fertilization.* Chapman and Hall, London.

Masui, Y. (1985) Meiotic arrest in animal oocytes. In: (Metz, C.B. and Monroy, A., eds.) *Biology of Fertilization.* Academic Press, New York, pp. 189–219.

Monroy, A. (1965) *Chemistry and Physiology of Fertilization.* Holt, Rinehart and Winston, New York.

Monroy, A., and Moscona, A. (1979) *An Introduction to Developmental Biology.* Chicago University Press, Chicago.

Ogielska, M. (ed.) (2009) *Reproduction of Amphibians.* Science Publisher, Enfield.

Schultz, R.M. (1999) Preimplantation embryo development. In: (Fauser, B.C.M., ed.) *Molecular Biology in Reproductive Medicine.* Parthenon Publishing, London, pp. 313–331.

Tarin, J., and Cano, A. (ed.) (2000) *Fertilization in Protozoa and Metazoan Animals.* Springer, Berlin.

Trounson, A., and Gosden, R. (eds.) (2003) *Biology and Pathology of the Oocyte.* Cambridge University Press, Cambridge.

Wassarman, P.M., Florman, H.M., and Greve, I.M. (1985) Receptor-mediated sperm-egg interactions in mammals. In: (Metz, C.B. and Monroy, A., eds.) *Fertilization*, vol. **2**. Academic Press, New York, pp. 341–360.

Wilson, E.B. (1900) *The Cell in Development and Inheritance.* Macmillan, London.

Yanagimachi, R. (1994) Mammalian fertilization. In: (Knobil, E. and Neill, J.D., eds.) *The Physiology of Reproduction.* Second edition. Raven Press, New York, pp. 189–317.

Selected Articles, Research Papers, and Reviews

Ajduk, A., Ilozue, T., Winsor, S. et al. (2011) Rhythmic actomyosin-driven contractions induced by sperm entry predict mammalian embryo viability. *Nature Communications.* DOI: 10.1038/ncomms1424.

Amundson, J., and Clapham, D. (1993) Calcium waves. *Current Opinion in Neurobiology* **3**:375–382.

Ao, A., Erickson, R.P., Winston, R.M.L. et al. (1994) Transcription of paternal human zygote as early as the pronucleate stage. *Zygote* **2**:281–287.

Aydin, H., Sultana, A., Li, S. et al. (2016) Molecular architecture of the human sperm IZUMO1 and egg JUNO fertilization complex. *Nature* **534**:562–565.

Bean, B.P. (1989) Classes of calcium channels in vertebrate cells. *Annual Reviews of Physiology* **51**:367–384.

Bedford, J.M. (1972) An electron microscopical study of sperm penetration into the rabbit egg after natural mating. *American Journal of Anatomy* **133**:213–254.

Bedford, J.M. (2014) Singular features of fertilization and their impact on the male reproductive system in eutherian mammals. *Reproduction* **147**:43–52.

Belton, R., Adams, N., and Foltz, K. (2001) Isolation and characterization of sea urchin egg lipid rafts and their possible function during fertilization. *Molecular Reproduction and Development* **59**:294–305.

Berridge, M.J. (1996) Regulation of calcium spiking in mammalian oocytes through a combination of inositol trisphosphate-dependent entry and release. *Molecular Human Reproduction* **2**:386–388.

Bianchi, E., Doe, B., Goulding, D. et al. (2014) Juno is the egg Izumo receptor and is essential for mammalian fertilization. *Nature* **508**:483–487.

Boveri, T. (1902) Über mehrpolige Mitosen als Mittel zur Analyse des Zellkerns. *Verh. Phys.-med. Ges. Würzberg N.F.* **35**:67–90.

Boveri, T. (1909) Uber Geschlectschromosomen Ober Nematoden. *Arch.F.ZellF* **4**:132–141.

Braude, P., Bolton, V., and Moore, S. (1988) Human gene expression first occurs between the four- and eight-cell stages of preimplantation development. *Nature* **332**:459–461.

Brownlee, C., and Dale, B. (1990) Temporal and spatial correlations of fertilization current, calcium waves and cytoplasmic contractions of eggs of Ciona Intestinalis. *Proceedings of the Royal Society B* **239**:321–328.

Carre, D., Rouviere, C., and Sardet, C. (1991) In vitro fertilization in ctenophores: sperm entry, mitosis and the establishment of bilateral symmetry in beroeovata. *Developmental Biology* **147**:381–391.

Claw, K., George, R., and Swanson, W. (2014) Detecting co-evolution in mammalian sperm-egg fusion proteins. *Molecular Reproduction and Development* **81**:531–538.

Clift, D., and Schuh, M. (2013) Re-starting life: fertilization and the transition from meiosis to mitosis. *Nature Reviews Molecular Cell Biology* **14**:549–562.

Collas, P., and Poccia, D. (1998) Remodeling the sperm nucleus into a male pronucleus at fertilization. *Theriogenology* **49**:67–81.

Costache, V., McDougall, A., and Dumollard, R. (2014) Cell cycle arrest and activation of development in marine invertebrate deuterostomes. *Biochemical and Biophysical Research Communications* **450**:1175–1181.

Dale, B. (2016) Achieving monospermy or preventing polyspermy. *Research and Reports in Biology* **7**:47–57.

Dale, B., and De Felice, L.J. (2011) Polyspermy prevention: facts and artifacts. *Journal of Assisted Reproduction and Genetics* **28**:199–207.

Dale, B., and Monroy, A. (1981) How is polyspermy prevented? *Gamete Research* **4**:151–169.

Dale, B., De Felice, L.J., and Ehrenstein, G. (1985) Injection of a soluble sperm fraction into sea urchin eggs triggers the cortical reaction. *Experientia* **41**:1068–1070.

Dale, B., De Felice, L.J., and Taglietti, V. (1978) Membrane noise and conductance increase during single spermatozoon-egg interactions. *Nature* **275**:217–219.

Dale, B., Gualtieri, R., Talevi R. et al. (1991) Intercellular communication in the early human embryo. *Molecular Reproduction Development* **29**:22–28.

Dale, B., Marino, M., and Wilding, M. (1998) Soluble sperm factor, factors or receptors. *Molecular Human Reproduction* **5**:1–4.

Dale, B., Tosti, E., and Iaccarino, M. (1995) Is the plasma membrane of the human oocyte reorganized following fertilisation and early cleavage? *Zygote* **3**:31–36.

Davidson, E.H. (1990) How embryos work: a comparative view of diverse modes of cell fate specification. *Development* **108**:365–389.

Dekel, N. (1996) Protein phosphorylation/dephosphorylation in the meiotic cell cycle of mammalian oocytes. *Reviews of Reproduction* **1**:82–88.

Dummollard, R., Carroll, J., Duchen, M. et al. (2009) Mitochondrial function and redox state in mammalian embryos. *Seminars in Cell and Developmental Biology.* **20**:346–353.

Eckelbarger, K. (2005) Oogenesis and oocytes. *Hydrobiologia* **535/536**:179–198.

Edwards, R. (1974) Human follicular fluid. *Journal of Reproduction and Fertility* **37**:189–219.

Edwards, R.G. (2004) New outlooks on gene activation and X-inactivation in mouse embryos. *Reproductive Biomedicine Online* **8**:248–250.

Edwards, R.G., and Beard, H. (1997) Oocyte polarity and cell determination in early mammalian embryos. *Molecular Human Reproduction* **3**:868–905.

Epel, D. (1990) The initiation of development at fertilization. *Cell Differentiation and Development* **29**:1–12.

Ernst, S. (1997) A century of sea urchin development. *American Zoologist* **37**:250–259.

Fleming, T.A., and Johnson, M.H. (1988) From egg to epithelium. *Annual Review of Cell Biology* **4**:459–485.

Fol, H. (1877) Sur le commencement de l'henogenie chez divers animaux. *Archives des Sciences Physiques et Naturelles Geneve* **6**:145–169.

Franzen, A. (1956) On spermiogenesis, morphology of the spermatozoon and biology of fertilization among invertebrates. *Zoologiska Bidrag, Fran Uppsala* **31**:355–482.

Fulka, J. Jr, First, N., and Moor, R.M. (1998) Nuclear and cytoplasmic determinants involved in the regulation of mammalian oocyte maturation. *Molecular Human Reproduction* **4**:41–49.

Galione, A. (1993) Cyclic ADP-ribose: a new way to control calcium. *Science* **259**:325–326.

Gianaroli, L., Tosti, E., Magli, C., Ferrarreti, A., and Dale, B. (1994) The fertilization current in the human oocyte. *Molecular Reproduction Development* **38**:209–214.

Groigno, L., and Whitaker, M. (1998) An anaphase calcium signal controls chromosome disjunction in early sea urchin embryos. *Cell* **92**:193–204.

Gualtieri, R., Santella, L., and Dale, B. (1992) Tight junctions and cavitation in the human pre-embryo. *Molecular Reproduction and Development* **32**:81–87.

Gurdon, J.B. (1967) On the origin and persistence of a cytoplasmic state inducing nuclear DNA synthesis in frog's eggs. *Proceedings of the National Academy of Science USA* **58**:545–552.

Hemmings, N., and Birkhead, T. (2015) Polyspermy in birds: sperm numbers and embryo survival. *Proceedings of the Royal Society B* **282**:1682–1688.

Hertwig, O. (1876) Beitrage zur Kenntniss der Bildung, Befruchtung und Theilung des thierischen Eies. *Morphologische Jahrbuch* **1**:347–434.

Holubcova, Z., Howard, G., and Schuh, M. (2013) Vescicles modulate an actin network for asymmetric spindle positioning. *Nature Cell Biology* **15**:937–947.

Horstadius, S. (1939) The mechanisms of sea urchin development, studied by operative methods. *Biological Reviews* **14**:132–179.

Hoshi, M., Moriyama, H., and Matsumoto, M. (2012) Structure of acrosome reaction inducing substance in the jelly coat of starfish eggs : a Mini review. *Biochemical and Biophysical Research Communications* **425**:595–598.

Inoue, N., Ikawa, M., Isotami, A. et al. (2005) The immunoglobin superfamily protein Izumo is required for sperm to fuse with eggs. *Nature* **434**:234–238.

Irniger, S. (2006) Preventing fatal destruction: inhibitors of the anaphase-promoting complex in meiosis. *Cell Cycle* **5**:405–415.

Jaffe, L.A. (1976) Fast block to polyspermy in sea urchin eggs is electrically mediated. *Nature* **261**:68–71.

Jaffe, L.F. (1980) Calcium explosions as triggers of development. *Annals of the New York Academy of Science of the USA* **339**:86–101.

Jaffe, L.F. (1983) Sources of calcium in egg activation: a review and hypothesis. *Developmental Biology* **99**:265–276.

Jaffe, L.F., and Creton, R. (1998) On the conservation of calcium wave speeds. *Cell Calcium* **24**:1–8.

Jamieson, B., and Rouse, G. (1989) The spermatozoa of the polychaeta (annelid): an ultrastructural review. *Biological Reviews* **64**:93–157.

Jamieson, B., Richards, K., Fleming, T. et al. (1983) Comparative morphometrics of oligochaete spermatozoa and egg acrosome correlation. *Gamete Research* **8**:149–169.

Johnson, M. (2009) From mouse egg to mouse embryo: polarities, axes and tissues. *Annual Review of Cell and Developmental Biology* **25**:483–512.

Jones, K.T. (2005) Mammalian egg activation: from Ca^{2+} spiking to cell cycle progression. *Reproduction* **130**:813–823.

Just, E.E. (1919) The fertilization reaction in *Echinarachnius parma*. 1. Cortical response of the egg to insemination. *Biological Bulletin* **36**:1–10.

Kashir, J., Deguchi, R., Jones, C. et al. (2013) Comparative biology of sperm factors and fertilization induced calcium signals across the animal kingdom. *Molecular Reproduction and Development* **80**:787–815.

Kawano, N., Yoshida, K., Miyado, K. et al. (2011) Lipid rafts: keys to sperm maturation, fertilization and early Embryogenesis. *Journal of Lipids*. DOI:10.1155/2011/264706.

Kidder, G.M., and McLachlin, J.R. (1985) Timing of transcription and protein synthesis underlying morphogenesis in preimplantation of mouse embryos. *Developmental Biology* **112**:265–275.

Kraft, M. (2013) Plasma membrane organization and function: moving past lipid rafts. *Molecular Biology of the Cell* **24**:2765–2768.

Levitan, D., and Peterson, C. (1995) Sperm limitation in the sea. *TREE* **10**:228–231.

Litscher, E., and Wassarman, P. (2014) Evolution and synthesis of vertebrate egg-coat proteins. *Trends in Developmental Biology* **8**:65–76.

Lupold, S.A., and Fitzpatrick, J. (2015) Sperm number trumps sperm size in mammalian ejaculate evolution. *Proceeding of the Royal Society of London B* **282**:2122–2130.

Means, A.R., and Dedman, J.R. (1980) Calmodulin – an intracellular calcium receptor. *Nature* **285**:73–77.

Menezo, Y., and Dale, B. (1995) Paternal contribution to successful embryogenesis. *Human Reproduction* **10**:1326–1327.

Miao, Y., Stein, P., Jefferson, W. et al. (2012) Calcium influx-mediated signaling is required for complete mouse egg activation. *Proceedings of the National Academy* **109**:4169–4174.

Monroy, A. (1965) An analysis of the process of activation of protein synthesis upon fertilization. *Archives Biology (Liège)* **76**:511–522.

Morgan, H., Santos, F., Green, K. et al. (2005) Epigenetic programming in mammals. *Human Molecular Genetics* **14**:47–58.

Moy, G., and Vacquier, V. (2008) Bindin genes of the pacific oyster Crassostrea gigas. *Gene* **423**:215–220.

Moy, G., Springer, S., Adams, S. et al. (2008) Extraordinary intraspecific diversity in oyster sperm bindin. *PNAS* **105**:1993–1998.

Nance, J. (2014) Getting to know your neighbour: cell polarization in early embryos. *Journal of Cell Biology* **206**:823–832.

Newport, G. (1853) On the impregnation of the ovum in the amphibian (Second Series, Revised). And on the direct agency of the spermatozoon. *Philosophical Transactions of the Royal Society of London* **143**:233–290.

Nurse, P. (1990) Universal control mechanisms resulting in the onset of M-phase. *Nature* **344**:503–508.

Oulhen, N., Reich, A., Wong, J. et al. (2013) Diversity in the fertilization envelope of echinoderms. *Evolution Development* **15**:28–40.

Parry, H., McDougall, A., and Whitaker, M. (2006) Endoplasmic reticulum generates calcium signalling microdomains around the nucleus and spindle in syncytial *Drosophila* embryos. *Biochemical Society Transactions* **34**:385–388.

Pincus, G., and Enzemann, E. (1935) The comparative behavior of mammalian eggs in vivo and in vitro: I. The activation of ovarian eggs. *Journal of Experimental Medicine* **62**:655–675.

Pochon-Masson, J. (1968) L'ultrastructure des spermatozoids vesiculaires chez les Crustaces Decapodes avant et au cour de leur devagination experimentale. *Annals Science Nazionale (Zool)* **10**:367–454.

Puppo, A., Chun, J.T., Gragnaniello, G., Garante, E., and Santella, L. (2008) Alteration of the cortical actin cytoskeleton deregulates Ca^{2+} signalling monospermic fertilization and sperm entry. *PLoS One* **3**:e3588.

Raj, I., Al Hosseini, H., Dioguardi, E. et al. (2017) Structural basis of egg coat-sperm recognition at fertilization. *Cell* **169**:1315–1326.

Ross, L., and Normark, B. (2015) Evolutionary problems in centrosome and centriole biology. *Journal of Evolutionary Biology* **28**:995–1004.

Rothschild, L. (1954) Polyspermy. *Quarterley Review of Biology* **29**: 332–342.

Rouse, G. (1992) Ultrastructure of spermiogenesis and spermatozoa of four Oriopsis species. *Zoologica Scripta* **21**:363–379.

Rouse, G. (2005) Annelid sperm and fertilization biology. *Hydrobiologia* **535/536**:167–178.

Rouse, G.W., and Jamieson, B.G.M. (1987) An ultrastructural study of the spermatozoa of the polychaetes Eurythoe complanata (Amphinomidae) Clymenella sp mand Micromaldane sp (Maldanidae) with definition of sperm types in relation to reproductive biology. *Journal of Submicroscopic Cytology* **19**:573–584.

Russo, G., Kyozuka, K., Antonazzo, L., Tosti, E., and Dale, B. (1996) Maturation promoting factor in ascidian oocytes is regulated by different intracellular signals at meiosis 1 and 11. *Development* **122**:1995–2003.

Sagata, N. (1996) Meiotic metaphase arrest in animal oocytes: its mechanisms and biological significance. *Trends in Cell Biology* **6**:22–28.

Santella, L., Alikani, M., Talansky, B. et al. (1992) Is the human oocyte plasma membrane polarised? *Human Reproduction* **7**:999–1003.

Santella, L., Limatola, N., and Chun, J. (2015) Calcium and actin in the saga of awakening oocytes. *Biochemical and Biophysical Research Communications* **460**:104–113.

Santella, L., Vasilev, F., and Chung, J.T. (2012) Fertilization in echinoderms. *Biochemical and Biophysical Research Communications* **425**:588–594.

Sardet, C., Paix, A., Prodon, F. et al. (2007) From oocyte to 16 cell stage: cytoplasmic and cortical re-organizations that pattern the Ascidian embryo. *Developmental Dynamics* **236**:1716–1731.

Saunders, C.M., Swann, K., and Lai, F.A. (2007) PLC zeta, a sperm-specific PLC and its potential role in fertilization. *Biochemical Society Symposia* **74**:23–36.

Schatten, G. (1994) The centrosome and its mode of inheritance: the reduction in the centrosome during gametogenesis and its' restoration during fertilisation. *Developmental Biology* **165**:299–335.

Shire, J.G. (1989) Unequal parental contributions: genomic imprinting in mammals. *New Biology* **1**:115–20.

Snook, R., Hosken, D., and Karr, T. (2011) The biology and evolution of polyspermy: insights from cellular and functional studies of sperm and centrosomal behaviour in the fertilized egg. *Reproduction* **142**:779–792.

Song, J.L., Wong, J.L. and Wessel, G.M. (2006) Oogenesis: single cell development and differentiation. *Developmental Biology* **300**:385–405.

Springer, S., Moy, G., Friend, D., Swanson, W., and Vacquier, V. (2008) Oyster sperm bindin is a combinatorial fucose lectin with remarkable intraspecies diversity. *International Journal of Developmental Biology* **52**:759–768.

Suzuki, N. (1995) Structure, function and biosynthesis of sperm-activating peptides and fucose sulphate glycoconjugate in the extracellular coat of sea urchin eggs. *Zoological Science* **12**:13–27.

Swann, K. (1990) A cytosolic sperm factor stimulates repetitive calcium increases and mimics fertilization in hamster eggs. *Development* **110**:1295–1302.

Talevi, R., Dale, B., and Campanella, C. (1985) Fertilization and activation potentials in Discoglossus pictus: a delayed response to activation by pricking. *Developmental Biology* **111**:316–323.

Torok, K., and Whitaker, M. (1994) Taking a long, hard look at calmodulin's warm embrace. *Bioessays* **16**:221–224.

Tosti, E. (1994) Sperm activation in species with external fertilization. *Zygote* **2**:359–361.

Tosti, E., and Menezo, Y. (2016) Gamete activation: basic knowledge and clinical applications. *Human Reproduction Update* **22**:420–439.

Tsien, R.W., and Tsien, R.Y. (1990) Calcium channels, stores and oscillations. *Annual Reviews of Cell Biology* **6**:715–760.

Vacquier, V. (1998) Evolution of gamete recognition proteins. *Science* **281**:1995–1998.

Van Blerkom, J. (2011) Mitochondrial competence in the human oocyte and embryo and their role in developmental competence. *Mitochondrion* **11**:797–813.

Van Blerkom, J., and Davis, P. (1995) Evolution of the sperm aster after microinjection of isolated human sperm centrosomes into meiotically mature human oocytes. *Human Reproduction* **10**:2179–2182.

Van Blerkom, J., Davis, P., and Lee, J. (1995) ATP contents of human oocytes and developmental potential and outcome after in vitro fertilisation and embryo transfer. *Human Reproduction* **10**:415–424.

Wade, P.A., Pruss, D., and Wolffe, A. (1997) Histone acetylation: chromatin in action. *Trends in Biochemical Science* **22**:128–132.

Walker, C., Harrington, L., Lesser, M. et al. (2005) Nutritive phagocyte incubation chambers provide a structural and nutritive microenvironment for germ cells of Strongylocentrotus droebachiensis, the green sea urchin. *Biological Bulletin* **209**:31–48.

Wassarman, P.M. (1990) Profile of a mammalian sperm receptor. *Development* **108**:1–17.

Wassarman, P.M. (1999) Fertilization in animals. *Developmental Genetics* **25**:83–86.

Wassarman, P.M. (2014) Sperm protein finds its mate. *Nature* **508**:466–467.

Wassarman, P.M., and Litscher, E. (2016) A bespoke coat for eggs: getting ready for fertilization. *Current Topics in Develpmental Biology* **117**:539–552.

Whitaker, M. (2006) Calcium at fertilization and in early development. *Physiological Reviews* **86**:25–88.

Whitaker, M. (2006) Calcium microdomains and cell cycle control. *Cell Calcium* **40**:585–592.

Wilding, M., Dale, B., Marino, M. et al. (2001) Mitochondrial aggregation patterns and activity in human oocytes and preimplantation embryos. *Human Reproduction* **16**:909–917.

Wilding, M., De Placido, G., Di Matteo, L. et al. (2003) Chaotic mosaicism in human preimplantation embryos is correlated with a low mitochondrial membrane potential. *Fertility and Sterility* **79**:340–346.

Wilding, M., Marino, M., Monfrecola, V. et al. (2000) Meiosis-associated calcium waves in ascidian oocytes are correlated with the position of the male centrosome. *Zygote* **8**:285–293.

Wilding, M., Russo, G.L., Galione, A. et al. (1998) ADP-ribose gates the fertilisation channel in ascidian oocytes. *American Journal of Physiology Cell Physiology* **275**:1277–1283.

Wilding, M., Wright, E.M., Patel, R. et al. (1996) Local perinuclear calcium signals associated with mitosis-entry in early sea urchin embryos. *Journal of Cell Biology* **135**:191–199.

Williams, C.J. (2002) Signalling mechanisms of mammalian oocyte activation. *Human Reproduction Update* **8**:313–321.

Yamasaki, M., Churchill, G.C., and Galione, A. (2005) Calcium signalling by nicotinic acid adenine dinucleotide phosphate (NAADP). *FEBS Journal* **272**:4598–4606.

Yanagimachi, R., Cherr, G., Matsubara, T. et al. (2013) Sperm attractant in the micropyle region of fish and insect eggs. *Biology of Reproduction* **88**:1–11.

Yanagimachi, R., Harumi, T., Matsubara, H., et al. (2017) Chemical and physical guidance of fish spermatozoa into the egg through the micropyle. *Biology of Reproduction*. DOI: 10.1093/biolre/iox015.

Index